Die Casting Metallurgy

Butterworths Monographs in Materials

The intention is to publish a series
of definitive monographs written
by internationally recognized
authorities in subjects at the
interface of the research interests
of the academic materials scientist
and the industrial materials
engineer.

Series editorial panel

M. Ashby FRS
University of Cambridge

J. Charles
University of Cambridge

A.G. Evans
University of California, Berkeley

M.C. Flemings
Massachusetts Institute of Technology

R.I. Jaffee
Electric Power Research Institute,
Palo Alto, California

R. Kiessling
Sveriges Mekanforbund, Stockholm

H. Suzuki
Tokyo Institute of Technology

I. Tamura
Kyoto University

G. Thomas
University of California, Berkeley

Forthcoming titles
Continuous casting of aluminium
Microorganisms and metal recovery
Control and analysis in steelmaking
Energy use in materials production
Amorphous metallic alloys
Residual stresses in metals
Eutectic solidification
Materials for plain bearings
Metallurgy of high speed steels
Energy dispersive X-ray analysis of
    materials
Mechanical properties of ceramics

Butterworths Monographs in Materials

Die Casting Metallurgy

Alan Kaye MSc, CEng, MIM
Chief Metallurgist, Fry's Diecastings Ltd, London

Arthur Street PhD, BSc

Illustrations by Rudi Hercik

Butterworth Scientific
London Boston Durban Singapore Sydney Toronto Wellington

This book is sold subject to the Standard
Conditions of Sale of Net Books and may not
be re-sold in the UK below the net price given
by the Publishers in their current price list.

First published, 1982

©Butterworths & Co (Publishers) Ltd., 1982

British Library Cataloguing in Publication Data

Kaye, Alan
  Die casting metallurgy. — (Butterworths monographs
  in materials)
  1. Die-casting
  I. Title   II. Street, Arthur
  671.2'53   TS239

  ISBN 0–408–10717–0

Typeset by Brechinset, Ipswich, Suffolk
Printed and bound in England by Butler & Tanner Ltd., London & Frome

Preface

The pressure die casting process is over 100 years old and during that time it has passed through many stages of development. It began as an offshoot of the printing industry; then its scope was widened, with alloys of increasing melting point, from lead to zinc, aluminium, magnesium and copper and finally to alloys of iron. During most of those years die casting development was dominated by designers and engineers. Consequently more emphasis was placed on die and casting design and manufacture than on the metallurgy of the die steels and casting alloys.

Metal values have always been a major part of the cost of any die casting, and it represents the largest single item of consumable stock. Even in the mid 1930s when zinc alloy die castings were sold at a now incredible 7d (about 3p or 6 cents) per pound, the raw material accounted for about half the cost of the average die casting. In spite of this, the influence of metal content on the production costs, the importance of metal losses and the possibilities of improving the efficiency of melting equipment did not begin to get adequate priority until the 1970s. Even now, metal losses are not measured with the same diligence as production speeds.

The impressive achievements of designers and engineers which led to the manufacture of dies for large and complex pressure die castings, such as automobile cylinder blocks, have been spectacular and well publicized. Less attention has been drawn to the metallurgy of die steels. The past decade has seen advances in the technology of die heat treatment but the die lives that are achieved are still far from satisfactory and the theory of die failure by thermal fatigue is so complex and almost beyond exact mathematical prediction that it is remote from most of those who are involved in die casting production.

Among many other skills, management require the ability to foresee what will happen in future years and to plan accordingly. Managers need to be aware of the availability of the metals which are die cast and the alloying elements which are used in die steels, all of which can depend on political considerations. For example, a crisis in Africa might prejudice not only the price and availability of copper, but also metals like chromium, manganese, vanadium and boron, without which the die casting industry would be hard pressed to make satisfactory dies.

There is a great deal of speculation that ore deposits of several essential metals will not be sufficient for man's needs in the near future. We have, therefore, given summaries of the ore distribution, smelting methods and energy requirements for the major non-ferrous metals which are die cast. The development of ferrous die casting, about which there are no qualms concerning metal availability but some about viability, is discussed in Chapter 17.

It is expected that a major part of the future increase in die casting production will be due to the widened use of aluminium in the automobile industry. Therefore it is important to discover what is being done in the development of smelting capability and in the manufacture of secondary alloys. Since metal scrap reclamation requires much less energy than the smelting of primary metals and since more efficient recycling of scrap may help to improve supplies of die casting alloys, we have included discussions of secondary metal refining, where relevant. Also, since silicon is an important but little publicized constituent of most aluminium casting alloys we have included a short chapter on its production.

It is sometimes difficult to correlate information about the properties of alloys because different specifications and units of measurement are employed in different parts of the world. For example, many die casters think of the strength of their alloys in terms of tons or pounds per square inch, while others use metric or SI units. Although attempts are being made in the Common Market to have one uniform set of alloy specifications, most die casters use the nomenclature that is familiar in their own countries. We have, therefore, attempted to summarize some comparative alloy specifications and to correlate units of hardness, strength and other properties.

The die casting literature, comprising a few books and many thousands of technical papers, has given much more emphasis to design, engineering and measurement than to metallurgy. We felt it would be useful to write a book on the metallurgical aspects of die casting, with discussions of some recent developments. In selecting the references shown at chapter ends we have covered a wide range of publications and have noted some papers which themselves give useful and extensive lists of references. A great deal of information has been obtained from papers presented to annual congresses of the Society of Die Casting Engineers, whom we thank for their cooperation. We also thank the American Die Casting Institute for their encouragement and help; we obtained many ideas from their report of the Tech Data 80 seminar.

We are grateful to many other friends in the die casting community, as well as those in supply and service industries, and in research and development establishments in the UK, America, Europe and Japan, who have generously given us the benefit of their advice in the preparation of this book.

A.K.
A.C.S.

Contents

Chapter 1

Early developments in die casting

The art of casting metals in permanent moulds can be traced to the dawn of civilization. Primitive man had learned to fashion and bake clay into useful objects, so clay moulds were used for making metal castings. In the early Bronze Age, the moulds were one piece and open topped; the castings produced from them were flat on one face and required hammering to make the final shapes. Later castings were made from two-piece moulds, giving a closer approach to the finished product; many such moulds for producing axe heads and other weapons are displayed in museums all over the world. By 3000 BC, the precious metals gold and silver were known; copper and bronze were fabricated and cast. Our metallurgical forefathers had discovered that copper is difficult to cast into complex shapes but, at a later stage in development, the metal was alloyed with tin to form bronze and it became possible to cast weapons, tools and ceremonial vessels of great beauty. Permanent metal moulds, the forerunners of die casting, began to be used over 3000 years

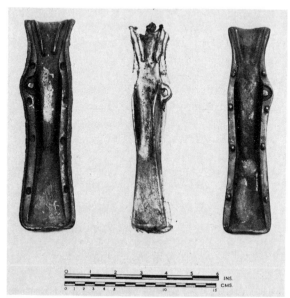

Figure 1.1 An ancient bronze two part mould and a palstave cast in it. (Courtesy Pitt Rivers Museum)

1

ago. *Figure 1.1* shows a two part bronze mould, dating to the second millenium BC, found near Mâcon in France. It has a series of dowels and sockets to locate the two halves; the pouring gate is flanked with two vents. In 1950, the British Non-Ferrous Metals Research Association used the mould to produce some castings in a 7% tin bronze, one of which is shown in the illustration. The castings are palstaves, a type of bronze axe head, produced during the Bronze Age. The french palstaves were of a particular kind, with a pair of flanges projecting from each face so that the axe head could be fitted into a wooden shaft. The castings and the ancient mould are exhibited in the Pitt Rivers Museum in Oxford and are described in their publication[1] on the prehistoric metallurgy of copper and bronze.

Mass production from metal moulds was developing over 2000 years ago. A six part bronze mould for casting three bronze arrowheads at a time, dating to the seventh century BC, was discovered at Mosul in Iraq[2]. This was a sophisticated piece of work, with an integral core, and was fashioned to make large numbers of arrowheads. In Britain about 25 bronze moulds have been found and there are some examples of clay or stone moulds with separate metal cores[3].

Lead was being cast over 5000 years ago, and many ancient castings in this metal have been discovered. Nearer our own time, the Romans used stone moulds[4] to mass produce pewter articles composed of 50% lead and 50% tin. Lead is easy to cast, but its weakness limited its use to decorative and votive objects. The excellent castability of the metal and the accurate reproduction which is possible, led to its use in the casting of printers' type and eventually to the development of pressure die casting.

In 1439, Johann Gutenberg invented a process for making printers' type-letters or a continuous text — from a lead alloy, cast into permanent moulds; thus each letter was identical with every other casting of the same letter. In 1540, Vanoccio Biringuccio[5], of Sienna, the author of the first book about metallurgy, described type-founding using an alloy with 83.85% lead, 12.3% tin and 3.85% antimony. He wrote. . . 'The impression of the letter you wish to cast is made by a steel punch on a little piece of copper. This hollowed out piece of copper is placed in the mould and the whole adjusted to receive the molten metal. When they have the desired quantity of one kind of letter they take out the matrix and insert another'.

For the next 300 years the techniques of typecasting continued to develop, until in 1822 William Church[6] introduced a machine with an output of up to 20 000 letters per day. This was the first successful use of die casting, although in 1805 a machine patented in the USA by William Wing had been abandoned after large sums had been spent on its development. The work of other pioneers, Bruce, Sturgis, Barr and Pelize, led to Otto Mergenthaler's production of a linotype casting in 1885, by means of a piston and a cylinder immersed in molten type-metal[7].

Thus, during the 19th century, the principles of pressure die casting were being applied to the specialized field of printing, and the process was beginning to extend to the production of other articles. One of the most extraordinary developments came from Charles Babbage who, among other achievements, invented an early computer containing many gears, cams, levers and cranks which had to be produced to a high degree of accuracy. Babbage made these by what he called 'pressure moulding in pewter hardened with zinc' in 1869. Some of these die castings are in the Science Museum in Kensington and they are described in a paper by H.K. Barton[6].

Those companies which supplied lead alloy to the printers, noticing the speed of output of lead type, must have wondered whether other castings could be produced by a similar process, as had already been done by Babbage. Some of the world's largest die casting companies owed their beginnings to the production of lead alloys for the printing industry, giving them the idea of extending the process in other directions.

In 1872 a small hand operated machine[8] was in operation. Molten metal was held in a container shaped like a kettle and, by striking a blow on a spring-supported knob, which depressed a plunger, the metal was forced up the 'spout' and into a mould. By the 1880s small lead and tin alloy components of cash registers and gramophones were being cast on such primitive machines. Bearings for automobile connecting rods were die cast in 1904 by the H.H. Franklin Company[8]. In that year the newly-born automotive industry replaced printing as the foremost user of die casting.

One of the first machines to produce die castings in zinc alloys as well as in lead and tin was patented by Herman H. Doehler in 1905 and this great pioneer[9], who had previously worked in the printing industry, created a company that is famous in die casting history. Doehler's machine is illustrated in *Figure 1.2*. A cylinder mounted in the holding

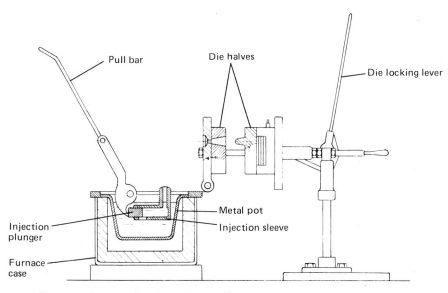

Figure 1.2 Doehler's first die casting machine

pot was submerged in molten metal; a plunger moved laterally in this cylinder, actuated by a pull bar and plunger arm. The molten metal filled the cylinder through an aperture. When the bar was pulled, the plunger moved forward, closing the filling aperture and then forcing the metal upwards. The two die halves were mounted on platens and opened and closed using a die locking lever, often requiring the efforts of two men. After the die was locked, the die and supporting frame were tilted manually to a position above the metal, to make contact with the filling nozzle. Operation of the pull bar, which forced molten metal into the die cavity, could often be done by one strong man, but large and complicated castings required the efforts of two 'pullers'. Retraction of the plunger uncovered a metal feeding hole in the injection sleeve to allow gravity filling of metal for the next casting. Following injection and solidification of the metal, the entire die and frame was levered into a horizontal position before the die halves were opened. Then the casting was pushed away by thin cylindrical rods — the forerunners of ejector pins.

It will be gathered from the description above that the first die casting machines required the strenuous manual effort of several operators but, as soon as the viability of the process was established, improvements were made to reduce the amount of labour and to extend its scope to the production of larger components. In 1907, van Wagner[8], introduced a machine which injected the metal by air pressure instead of by pull bar but the die still had to be tilted through 90 degrees, as with Doehler's early machine, and the 'gooseneck' injection chamber had to be filled by hand ladling. In 1914 Doehler introduced a machine in which the die was moved horizontally beneath the nozzle seat; air pressure applied to the top surface of molten metal in the holding pot forced an appropriate amount of molten metal up a channel into the die, and such machines proved capable of die casting aluminium alloys. In his book[10], Doehler stated that during World War One the process was sufficiently developed to produce aluminium components of machine guns, gas masks and binoculars.

Metallurgical problems in the industry had become apparent early in the century. Doehler and his customers found that zinc die castings were deteriorating in service. Realizing that contamination by lead and tin was causing the trouble he removed those metals to a separate part of his plant. He also embarked on the casting of aluminium as an alternative to zinc. In 1907, one of Doehler's consultants made the diagnosis that even a trace of lead will weaken zinc alloy and he went on to remark, prophetically, 'Some day zinc producers will find a way to produce pure zinc'[11].

Early machines required enough molten metal for each shot to be ladled into the injection chamber prior to casting, but in 1920 the Madison Kipp Corporation of the USA introduced a pneumatic 'hot chamber' machine where the injection chamber contained enough metal for many cycles. Between 1926 and 1928 the same company produced a mechanized machine, shown in *Figure 1.3*, operated by a chain drive from the motor to the shaft and then through worm and wormwheel to a cam shaft. As the cams rotated, connecting rods reciprocated, moving the carriage back and forth. During the forward

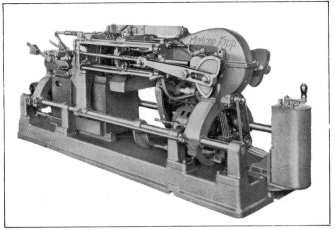

Figure 1.3 An early mechanized hot chamber machine. (Courtesy Madison Kipp Corporation)

movement the gooseneck was brought up to the casting position after the die halves were closed. With further rotation of the camshaft, a lever opened an air valve which allowed compressed air to force metal held in the gooseneck into the die cavity; the inlet air valve was closed and the exhaust air valve opened, allowing the entrapped air to escape; finally the moveable die was opened and the casting ejected.

At that time, most die castings were produced in alloys of lead or zinc. Low melting point and easy castability were paramount requirements and alloys of zinc with tin proved to be suitable, with the additional advantage of ease of soldering. Later, zinc-aluminium—copper alloys were developed but, even in the late 1920s, when the deleterious effect of lead and tin on the properties of these new alloys was being revealed, die casters still had low-grade zinc—aluminium alloys circulating in their foundries. The practice of adding tin or lead to aid fluidity was an easy but dangerous way of helping production and did not cease until it was realized that the zinc—aluminium alloys, with or without copper, must be based on super-pure zinc free from the harmful impurities.

Another early foundry practice was the use of 'secret' compositions for die lubricants, each skilled operator having his own mixtures. It is recollected that in the same establishment which produced the first experimental cylinder block, a die lubricant named 'mayonnaise' was used with great confidence. When efforts were made to mechanize the process in the 1950s, one of the hardest problems to be overcome was the reluctance of operators to cease using their own favourite lubricants.

Early in the 20th century, aluminium alloys were being cast in sand and gravity casting permanent moulds, but the higher melting point (about $600°C$ compared with $400°C$ for zinc alloys) introduced problems in pressure die casting. At first efforts were made to use the submerged plunger air pressure machines that had been successful for zinc alloys but the higher temperature and the tendency of molten aluminium alloy to attack the iron and steel parts of the machine limited the scope of the hot chamber process. In 1920, a significant development was made in the invention of the 'cold chamber machine', built by Carl Roehri[11]. The alloy was melted separately and hand ladled into an injection sleeve. Immediately after pouring, a plunger forced molten metal into the die; then the casting was ejected with the excess metal attached to it in the form of a cylindrical slug.

Two types of cold chamber machine were developed. The vertical injection machine was invented by Josef Polak of Prague in 1927, originally with the intention of die casting brass, but the same equipment was found to be suitable for aluminium[12]. *Figure 1.4* illustrates the operating principles of this machine in which the pressure was generated by water-hydraulic systems. Metal sufficient for one shot was ladled into a cylinder, of which the bottom portion was a piston held in position by a spring. Then an upper piston moved downwards, thus depressing the molten metal and the piston on which it was resting. The bottom piston was forced past an orifice leading to the die cavity and the molten metal was injected through that orifice. The upper piston was retracted and the lower one returned upwards by the power of the spring; this cropped the slug of surplus metal away from the casting gate. The die was opened, the casting was removed and the slug returned to the melting pot or to a remelt furnace. The configuration of the Polak machine made it particularly suitable for casting designs that could be centre-run.

The other cold chamber process, developed at about the same time, involved a horizontally acting plunger, as shown in *Figure 1.5*. Here the slug of excess metal was

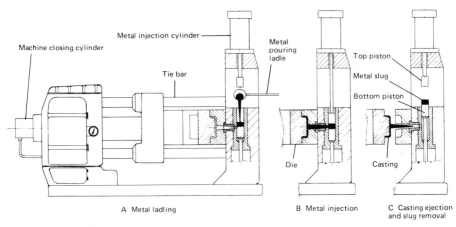

Figure 1.4 Vertical injection Polak cold chamber machine

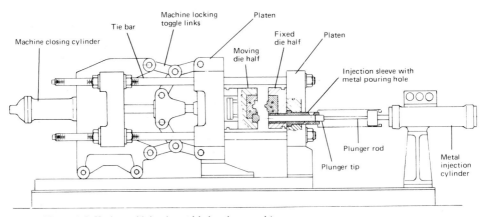

Figure 1.5 Horizontal injection cold chamber machine

ejected from the die in one piece with the casting. This was suitable for casting designs that could be side-run as shown in the illustration.

The vertical injection machines had advantages and disadvantages. When molten metal was ladled into the injection chamber, any oxide would float on the metal and, being the last to be injected, would remain in the slug of surplus metal attached to the die casting. In the horizontal machine the molten metal, before injection, lies in the sleeve, from where it is forced into the die by the plunger, along with any oxide or dross which would thus be included in the die casting itself. This was a disadvantage in the early years, when the movement of the injection plunger of the horizontal machine was not regulated, but even then horizontal machines had the advantage of being less wasteful in maintenance time.

The vertical machine system of a floating piston, on which the molten metal rested temporarily, caused metal to become entrapped between the pistons and the cylinder wall, leading to production delays and piston replacements. When, in the mid 1960s, two stage and three stage injection was developed, the horizontal machine came to produce as good if not better quality castings than the vertical ones and the machines were more reliable.

Early cold chamber machines were operated by emulsions of oil in water systems and even at present there are some plants operating successfully in this way. Water has the advantage of low cost and freedom from fire risk but its low viscosity causes leaks to be an ever present problem. The emulsion has to be circulated from a central pump station, provided with stand-by pumps in case of breakdown. Oil hydraulic systems are self contained; the fluid is more leak proof than water and, since the development of fire resisting fluids, oil hydraulic machines have become safe.

The modern advance into oil hydraulic machine operation was initiated by the invention of the Williams–Janey pump in the USA, first patented in 1907. However, it was not until 1926 that the idea of a self contained hydraulic system was introduced and such a unit was fitted into a hydraulic press. At about this time, work in the USA by Harry Vickers was concerned with self contained hydraulic power systems used in conjunction with gas and oil pressure accumulators and many such systems were incorporated in die casting machines of the 1940s. This opened the way for high pressure locking systems and the advanced characteristics which are commonplace in machines of the 80s.

Air operated machines for casting low melting point alloys developed in two directions. Some of them applied air pressure directly on the face of the molten metal in the inherent metal container, to inject the required amount of metal into the die; this mechanism was also used for the early pressure die casting of aluminium. In order to give better control to the amount of metal injected, to avoid the deterioration of metal quality caused by the air in contact with the metal and to achieve higher injection pressures, air operated plungers were developed. Among other pioneers , Jack Schultz in the USA, built machines from 1924 onwards that used this principle for metal injection, while die closing and locking were effected by hydraulic cylinders, often associated with mechanical toggle links. In the UK, hot chamber machines were introduced by the EMB Company employing air operated plungers and die closing mechanisms. These machines were developed into fully mechanized units which were among the first in the field of die casting automation.

Die locking systems have developed round toggle link mechanisms whereby the two halves of the die attached to the machine platens are held together by mechanical links and tie bars. Much importance is now centred in embodying a rigid locking structure which provides accuracy in die alignment and satisfies the demands for increasing injection pressures. Distances between machine platens must be variable, to accommodate different die thicknesses and allow for thermal expansion of the dies. On large machines, where adjustment of locking nuts would be arduous and time wasting, a gearing system is now used to ensure equal tension on all tie bars. *Figure 1.2* on page 3 should be compared with the machines illustrated on page 175 and elsewhere, to show the immense changes that have occurred during the past 80 years.

Although in the early years of the industry it was admitted that die casting was more of an art than a science, there were signs that research on the variables of the process were opening the way for the developments in instrumentation and the mathematical study of metal injection which have made such dramatic changes during the past 20 years. Hiram K. Barton's review 'The Pressure Diecasting of Metals'[13] paid tribute to the prewar work of Dr Leopold Frommer and the postwar researches of Professor Seita Sakui and his coworkers at the Tokyo Institute of Technology.

In the 1950s very large die casting machines were developed, leading to the American production of radiator grilles in zinc alloy and the first experimental cylinder blocks in aluminium alloy, described on page 31.These opened up many new fields for the industry, but emphasized that manual operators could not ladle such heavy amounts of molten metal unassisted. Many ingenious devices were invented for mechanizing molten metal transfer and for removing the casting from the die opening. During the following years automatic metal pouring, die and machine lubrication, robots for handling the castings and progressive developments in instrumentation[14] led to the situation where the industry requires sophisticated technical skills and the availability of capital on a scale which would have appeared unbelievable to the early die casters.

The technical developments which have been outlined in this chapter emphasized the work of countless engineers — enthusiastic for the possibilities of pressure die casting — who have made the process possible. During recent years there has been a concerted effort to standardize the construction and classification of die casting machines[15]. In Britain the new proposed standards have been developed by the Diecasting Society, the Light Metal Founders' Association and the Zinc Alloy Die Casters Association. In harmony with their efforts, the work of the component designer has been vital. One has only to

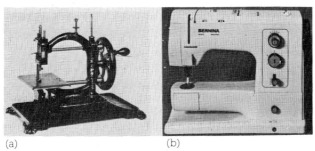

(a)   (b)

Figure 1.6 (a) An early sewing machine produced in cast iron. (b) A modern sewing machine incorporating zig-zag and other fancy stitches. Produced from pressure die castings. (Courtesy Bernina Museum and Bernina Fritz Gegauf Ltd., Switzerland)

compare the picture of an early sewing machine with that of a modern one, both shown in *Figure 1.6,* to appreciate how 'designing for die casting' has widened the scope of the process and has benefitted manufacturers who apply die casting to their products.

While this engineering progress has been continuing, there have been many metallurgical developments in the die casting industry, although these have not been spectacular in the literal sense of the word. However, the past 30 years have seen an increasing realization that die casting requires a combination of metallurgy with mechanical and electrical engineering; the Authors are convinced that the coming decade will be one in which metallurgical progress is at least as important as engineering. Metal accounts for about one-third of the production cost of a typical aluminium die casting weighing about 1 kg; metal melting is about 7% of the total cost. Metal losses and die heat treatment are two other quite substantial factors in the make up of production cost. Consequently it is expected that improvements in the melting efficiency of furnaces, the control of melting to reduce metal losses, and further studies of the selection, heat treatment and behaviour in service of die steels will contribute even more significantly than further increases in production rate.

References

1. COGHLAN, H.H.
    Notes on the Prehistoric Metallurgy of Copper and Bronze in the Old World. *Pitt Rivers Museum, Oxford, Occasional Papers on Technology* Vol. 4, 2nd edition p. 59 and plate IX (1975)

2. MARYON, E.
    Metal Working in the Ancient World. *American Journal of Archeology;* **53,** 93 (1949); **65,** 173 (1961)

3. MOHEN, J.P.
    L'age du Bronze dans la région de Paris. *Edition des Musées Nationaux Paris* p. 540–546 (1977)

4. TYLECOTE, R.F.
    A History of Metallurgy. The Metals Society p. 62 (1976)

5. BIRINGUCCIO, V.
    De la pirotechia, Venice (1540) English translation of *Typecasting in the 16th Century*. M.T. Gnudi and C.N. Smith, Columbia Club of Connecticut, New Haven (1941)

6. BARTON, H.K.
    Charles Babbage and the beginnings of die casting. *Die Casting Engineer,* **20,** (4), 12 (July/August 1976)

7. STERN, M.
    Die Casting Practice. McGraw–Hill Book Company, p. 162 (1930)

8. (Editorial)
    The Golden Age of Die Casting. *Die Casting Engineer,* **20,** (4), 25 (July/August 1976)

9. (Editorial)
    Some Pioneers in the U.S. Die Casting Industry. *Die Casting Engineer,* **20,** (4), 34 (July/August 1976)

10. DOEHLER, H.H.
    Die Casting McGraw–Hill Book Company, New York, p. 2 (1951)

11. (Editorial)
    The Custom Die Caster. *Precision Metal Moulding,* p. 50 (May 1964)

12. KAYE, A.
    Diecasting – 75 years of Development. *Foundry Trade Journal,* p. 390 (August 18th 1977)

13. BARTON, H.K.
    The Pressure Diecasting of Metals. *Metallurgical Reviews, Institute of Metals,* **9,** 36 (1964) (1964)

14. STREET, A.C.
    Developments in Pressure Diecasting. *International Metallurgical Reviews* The Metals Society and the American Society for Metals Vol. 20 (1975)

15. *Diecasting Society News London,* **3,** 20–35 (June 1981)

Chapter 2

Die casting metals and alloys

Apart from pure aluminium, which is die cast around laminated units in the manufacture of rotors for electric motors, described on page 39, pressure die castings are made of alloys. Compared with pure metals, alloys are stronger; for example the aluminium alloy with about 11% silicon is more than twice as strong as aluminium. An alloy has a finer grain size and greater strength when rapidly cooled than when slowly cooled. The average grain size of a pressure die casting (about 0.01 mm) is smaller than that of a permanent mould casting (about 0.5–1.0 mm) and a sand casting (1.0 mm and upwards). Thus a pressure die casting will be stronger than a slowly cooled permanent mould casting of the same alloy and that in turn will be stronger than a sand casting. There is no need for grain refining or modification treatments in pressure die casting, as is sometimes required in other casting methods.

When two metals which do not exhibit complete solid solubility are alloyed together, there is usually a eutectic composition which has a freezing point lower than that of either of the constituents. In the aluminium-rich alloys with silicon, the eutectic appears at about 12% silicon and the melting point of the alloy is nearly 100°C lower than that of aluminium and about 850°C lower than silicon. *Table 2.1* shows the eutectics in some alloy systems used in die casting.

TABLE 2.1

Alloy System	Eutectic composition	Eutectic temperature (°C)
Aluminium–silicon	11.7% Si*	577
Aluminium–magnesium	35.0% Mg	450
Zinc–aluminium	5.0% Al	382
Magnesium–aluminium	32.3% Al	437
Lead–antimony	11.2% Sb	252

* The composition and melting point of the aluminium—silicon eutectic varies slightly according to the rapidity of cooling and whether or not the alloy is modifed, as discussed on page 23.

On solidification, binary eutectic alloys have characteristic structures consisting of an intimate mechanical mixture of the two phases. If such alloys possess other properties which make them suitable for die casting, they are obvious choices for the process, because their lower melting points will lead to longer die lives than would be obtained with alloys of higher melting points.

Most metals are soluble in each other when liquid, and many of them can retain another metal in a state of solid solution. A few pairs of metals, such as the copper—nickel alloys, are solid soluble within the whole range of composition. The solid solubility of one metal in another is greatest near the melting point and decreases at diminishing temperatures. For example, at 380°C, just below the eutectic temperature of zinc with aluminium, the solid solubility of aluminium is about 1% but this decreases to about 0.5% at 250°C and to 0.05% at room temperature. In contrast to pure metals and eutectics, solid solution alloys melt and freeze over a range of temperature. Their freezing ranges encourage three separate zones to be formed during solidification: a zone of solid metal adjacent to the die face, a liquid zone surrounding the heat centre of the casting and an intermediate pasty region of solid and liquid. Producers of permanent mould and sand castings can experience some difficulty in making sound castings in alloys of this type. Feeding of liquid metal through the intermediate region is restrained by the pasty solid/liquid mixture; the tendency for trapped porosity increases with alloys of longer freezing range. In pressure die casting of such alloys, for example, the magnesium—aluminium alloys, the temperature gradients during solidification are higher than in permanent mould and sand casting, and the intermediate pasty region will be correspondingly narrow. The high pressure applied in the final stage of injection assists in feeding the casting.

Table 2.2 shows the liquidus melting points and solidus freezing points of the principal die casting alloys. Since all of them contain permitted amounts of impurities, the liquidus and solidus temperatures are slightly different from those measured in completely pure alloys.

TABLE 2.2

Alloy System	Typical composition	Liquidus (°C)	Solidus (°C)	Freezing range (°C)
Aluminium—silicon	LM6 (413.2) 11% Si	575	565	10
Aluminium—silicon—copper	LM24 (380.0) 4% Cu 9% Si	580	520	60
Aluminium—magnesium	LM5 (514.0) 5% Mg	642	580	62
Zinc—aluminium	BS1004A (AG40A) 4% Al	387	382	5
Magnesium—aluminium	MAG7(AZ91B) 9% Al	596	468	128
Lead—antimony	5% Sb	310	270	40

The latent heat of fusion (heat/unit mass) that is released when a metal changes from liquid to solid occurs at a constant temperature for pure metals and eutectics as indicated by the horizontal lines in *Figure 2.1* but is released over the solidification range for alloys. Latent heat of fusion for the aluminium alloy LM24 is about four times greater than for zinc alloy and 1.7 times greater on a volume basis. Zinc alloy solidifies more rapidly than aluminium alloy and requires faster injection speeds to allow the rapid transmission of injection pressure to the entire section of the solidifying casting.

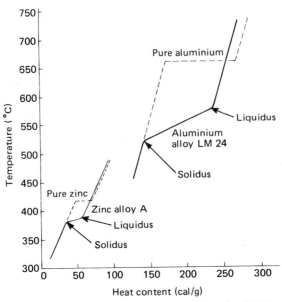

Figure 2.1 Heat content curves for aluminium alloy LM24 and zinc alloy A

The pressure die cast alloy is held in a container at a temperature just above its melting point and then injected into the die cavity at a speed of 20–50 m/s, depending on the characteristics of the alloy and the surface finish required on the castings. As the first part of the metal touches the die walls it freezes; within a fraction of a second the die cavity is full and the alloy, which a moment ago was liquid, is being consolidated into a casting with the assistance of intensified pressure. A few seconds later the hot die is opened, the casting is pushed from the moving die face by ejectors and falls either into a bath of water, on to a conveyor, or is grabbed by a robot.

The time scale of such a series of events may be faster or slower depending on the weight of the casting but the transition from molten metal to solidified casting involves rapid dimensional changes and hence considerable stresses. Thus it is essential that a pressure die cast alloy must be strong at a temperature just below that of the solidus. Alloys which do not possess sufficient strength at those temperatures are termed hot short and they tend to crack when solidifying in the die. This undoubtedly limits the range of pressure die cast alloys. For example 70/30 brass, an otherwise excellent material, is hot short and, therefore, unsuitable for die casting. The aluminium alloy with 10% magnesium is sand cast and can be permanent mould cast, providing greater strength than that obtained in alloys with less magnesium, but because the 10% alloy is hot short it cannot be pressure die cast except in the simplest of shapes. As a third example the magnesium content in zinc alloy die castings is restricted within the limits 0.03–0.06%. Increase of magnesium above 0.06% causes the alloy to be hot short.

The choice of alloy

Recent economic and industrial developments have affected the choice of alloys to be used for present production and for future capital investment. Raw material price, as always, has been a powerful factor in deciding which alloy shall be used, but the price of a metal depends on the availability of its ores and the amount of energy required to convert it from ore to metal.

Technological changes within the industry also affect the choice of material; for example the development of thin-wall technique for zinc alloy die casting has made that metal increasingly competitive, and improvements in hot chamber machines for magnesium have helped to reduce costs, especially for small and medium sized components. The first decisions on choice of an alloy are made by the users, who know the conditions under which their products will operate but they are helped by the advice provided by the die casting industry on the subject of availability, price, design and finishing processes. Any die casting management which is prepared to spend large sums in capital investment needs to be aware of the future availability of the metals which are used, in order to avoid investing money in production which may decline owing to material shortages or steep price increases.

Although metals are purchased by the tonne, they are manufactured into articles of a given volume. Thus, a tonne of lead alloy would make 1000 components each weighing 1 kg but a tonne of magnesium alloy would make 6000 components of the same size. In the electrical industry, the choice of aluminium versus copper is related to the fact that, although aluminium has only about two-thirds the electrical conductivity of copper, a tonne of aluminium has more than three times the volume of copper and is about half the cost. Consequently from the point of view of cost per unit property in an electric conductor, aluminium is preferred.

A significant 'new look' on the choice of material was stimulated by Professor W.O. Alexander[1,7] and his colleagues at the University of Aston, causing designers and users of components to consider the 'cost per unit strength' of competing materials. The section thickness of an article is determined by the stress it has to endure and the ability of the material to withstand that stress. Thus a stronger material can be made thinner than the part it replaces — or a weaker material must be made thicker. This concept has led to the realization that despite the apparent lower costs of plastic injection mouldings, they are not necessarily a good bargain because their cost per unit strength is less favourable than would appear from their cost per tonne. Some further disadvantages of plastics are becoming apparent: they are made from oil products; furthermore they require the expenditure of considerable amounts of energy in their manufacture. Recycling of plastic materials is often difficult, sometimes impossible. Thus the competition from plastics, which fifteen years ago was thought a menace to the die casting industry, has become less powerful.

The total amount of energy required to provide a component of the required strength is a relationship of even greater relevance than cost per unit property[2]. Those who are planning for the future now think in terms of the amount of energy required for alternative materials to achieve the required properties in a given component. There is still some confusion between the energy required for the final smelting process (for example converting aluminium oxide to aluminium in electrolytic cells) and the grand total, including mining the ore, its concentration near the mine, transport to the smelting plant, pretreatment to prepare it for smelting and the final production of the pure metal. The differences between the two areas of measurement are considerable. For example a tonne of aluminium requires about 15 000 kWh for the final electrolytic separation, but when all the stages from ore at the mine to ingot at the die casting plant are added together, the total energy requirement is over 60 000 kWh per tonne of metal.

To allow a proper comparison, all of the forms of energy should be expressed in one standard, say kWh. This involves converting gas and oil to equivalent amounts of electrical energy. The fuel required for land and sea transport, the power used in mechanical equipment such as conveyors and crushers, the energy required in concentration and pretreatment of the ore must all be converted to the same units to make any comparison valid.

In the past, users of die castings tended to specify alloys which had been recommended for specific purposes or which appeared to be suitable from literature searches. Managements of 20 years ago were often willing to embark on the die casting of small batches of special alloys. When the costs of such operations were investigated it was found that special materials required considerable price increases. The cost of separate storage, separate melting, problems of separating castings, runners and scrap all added considerably to production cost. At present most die casting companies limit the range of alloys to a reasonable minimum, two or three aluminium alloys and one of zinc.

The new technology of computer aided die design and the growing acceptance of the 'thin-wall zinc' concept has led to zinc alloy die castings, which were declining in use in the late 1970s, coming again into healthy competition with aluminium and plastics. The thin-wall concept has, however, caused one part of the industry's equipment to decline in use. In the 1960s, before raw material prices had escalated through inflation, large and heavy zinc alloy die castings such as automobile radiator grilles were manufactured in hot chamber machines. This type of production declined as heavy zinc castings competed less favourably, and many hot chamber machines went out of use. At present the average shot weight in a hot chamber machine is less than before, partly because the thin-wall casting is lighter and partly because the new technology has reduced the weight of the gate and other surplus metal to about half of what it was a few years ago. All of these developments have caused small and medium sized hot chamber machines to be required, while very large ones are used to a much lesser extent.

Metal availability in the future may not appear such a pressing problem as the need to find enough orders to make a die casting plant run profitably. Nevertheless, during the postwar years much attention has been given to the future availability of metals. In 1952 the Paley Commission reporting to the President of the USA prophesied that the world's lead and zinc reserves would be exhausted by 1970, yet in that year, the reserves stood higher than ever before. Several more recent publications quote data suggesting that the current reserves of some metals are not adequate to last till the end of the present century[3,4,6]. However, although such predictions have proved unduly pessimistic in the past, the fact remains that the sources of metals are becoming depleted, and as their richness declines the amount of energy needed to extract the metals increases. New deposits are certainly being discovered but they are often in remote parts of the world, so that the cost of transport becomes a large factor in the total cost of the metal. It is, therefore, advisable that die casting managements make themselves aware of the distribution of the metals which they use and the possibilities of shortages or steep cost increases in the future.

Aluminium is widely distributed but requires large amounts of energy[7], not only for the final electrolytic extraction of the metal but also for all the preliminary stages of transport and refining of the bauxite ore to pure alumina. Zinc requires less energy than aluminium but its ores are weak and its distribution less widespread. Therefore the future prospects for zinc will depend partly on the discovery of fresh deposits (as has been done in Eire and Greenland) and improved extraction processes. Magnesium is available in limitless amounts since it is present in sea water, but the amount of energy required for its extraction is comparable with that for aluminium. Nevertheless, on the grounds of future availability no die casting management should ignore the potential for magnesium[5].

Copper alloys are necessary for many electrical uses and possess good resistance to corrosion; brass will, therefore, continue to have a place in the repertoire of die casters. Copper is one of the metals likely to be in short supply at the end of this century, but new methods of extracting the metal from weak ores and residues are being developed.

Lead ores are weak, but they require comparatively less energy to extract than for other die cast metals. The future of transport will be related to the developments of battery-powered vehicles, so the demand for lead will probably increase as this form of motive power develops.

There has been a great deal of research to develop new alloys that will provide increased strength and corrosion resistance, combined with good diecastability and economic cost. Although this has led to the marketing of some new materials, the basic repertoire of the die casting producer has been practically unchanged for at least 20 years. There are many possibilities of metallurgical research in problems associated with die casting. The Authors suggest two fields which may be interesting. Firstly, the conditions under which a molten alloy is injected into a die are quite different from those where an alloy has been cast and cooled more slowly. The grain structure of a pressure die casting shows the effect of high pressure injection followed by rapid cooling. Further studies of the structures of die cast alloys at high magnifications and with the resources of modern metallography would probably throw useful light on the behaviour of die cast alloys.

Secondly, there is a need for unification in the measurement of strength of die cast specimens. Manufacturers often have to provide guidance to their customers on the relative strengths of various die cast alloys and it is notorious that published figures vary widely because of the differences in measured strength according to the section thickness of the casting or test piece. As an extreme example the Authors found one reference where the tensile strength of the magnesium alloy MAG 7 was stated as $125 \, \text{N/mm}^2$ while in another reference the strength of the same composition given the designation AZ 91 was stated as $200-250 \, \text{N/mm}^2$. The 0.2% proof strength of the first alloy was given as $85 \, \text{N/mm}^2$ while that of the similar alloy was $150-170 \, \text{N/mm}^2$.

It is well known that the outer skin of a die casting is substantially stronger than the interior. Indeed the strength of a 'thin-wall' casting, which will be discussed on page 103 is maximized because all of the material in such a thin section is fine grained. When strength figures for die cast alloys are quoted they are often qualified by remarks about the difference between the strength of a thin section die casting and that of a conventional tensile test piece. Probably the best way out of the dilemma would be to provide sets of test figures for each alloy, linked with a range of section thicknesses, or better still with the grain size of the castings.

References

1.   ALEXANDER, W.O.
        Energy and Material. *The Metallurgist and Materials Technologist*, **12**, (9), 505 (September 1980)

2.   ALEXANDER, W.O.
      Total energy content of some metals and materials and their properties. *The Metals Society, London,* p.21–29 (May 1979) (part of book 265 published in 1980 by the Metals Society)

3.   DOWSING, R.J.
      Future Metal Strategy; Part 1. *Metals and Materials,* p. 19 (June 1979)

4.   DOWSING, R.J.
      Future Metal Strategy; Part 2. *Metals and Materials,* p.43 (September 1979)

5.   PYE, A.
      Substitution makes its mark. *Industrial Purchasing News,* p. 34a (May 1981)

6.   NATO SCIENCE COMMITTEE STUDY GROUP
      Rational Use of Potentially Scarce Metals. NATO Scientific Affairs Division (1976)

7.   ALEXANDER, W.O.; APPOO, P.M.
      Material selection, the total concept. *Design Engineering,* pp.59–66 (November 1977)

Chapter 3

The production of aluminium and its alloys

Small amounts of aluminium had been extracted by chemical processes early in the 19th century, and in 1886 the way was opened for the industrial production of the metal. In that year Charles Hall and Paul Héroult independently discovered that about 5% of aluminium oxide would dissolve in molten cryolite (sodium–aluminium–fluoride) at a temperature of under 1000°C, and that a powerful direct current passed through the mixture would separate the oxide into aluminium and oxygen.

As often happens, the fruitfulness of one invention was made possible by the timing of other developments. The theory on which the working of the dynamo was to be based had been discovered in the 1830s; some of the technical problems in the construction of a dynamo had been overcome in the 1860s and by 1879 it had become a viable machine. In 1890, following Hall and Héroult's work, Karl Josef Bayer invented the process for separating pure aluminium oxide from bauxite by the use of sodium hydroxide under pressure. These metallurgical and electrical inventions paved the way for the bulk production of aluminium, which today stands at the head of all non-ferrous metals in tonnage and industrial importance and which represents the major part of the die casting industry's output.

Aluminium is the most plentiful metal in the earth's crust[1]. It appears in many minerals, ranging from the oxide in bauxite to complex compounds of the metal with oxygen, silicon, calcium, magnesium and iron in clays. All aluminium-bearing minerals are difficult and costly to smelt but the aluminium oxide in bauxite provides the most economical source of the metal. The largest amount of bauxite is now mined by Australia, which extracts sufficient ore for half the world's aluminium output of about 13 million tonnes/year. Jamaica and Guinea come next and the USSR fourth. Thus the free world, the communist world and the third world each have a large stake in the mining of aluminium ore. That is the present position but, for information about the future potential for bauxite mining, the reserves (many of them undeveloped) have to be considered. Although the bauxite reserves in Australia are probably the largest, estimated at about 5200 million tonnes, there are vast deposits in Guinea (5000 million tonnes) and Brazil (2100 million tonnes)[2]. It is also estimated that Jamaica, Mali, India, Cameroon, Surinam and Guyana each hold bauxite reserves of around 1200 million tonnes. These major reserves total about 20000 million tonnes, to which smaller amounts elsewhere must be added. Taking a cautious estimate of 5 tonnes of bauxite per tonne of aluminium, there would be sufficient ore for about 300 years, at the present level of production. If one allows for an increase of 5% per annum in the consumption of aluminium, the reserves would last about 70 years. The major problems to be overcome in the immediate future arise from shortage of energy, not of bauxite.

The prospect for future supplies of aluminium ores appears secure and when, eventually, the richest bauxite deposits are worked out, the clays may provide a source of aluminium for many centuries to come, although the technical problems of extracting aluminium from clay are more involved[3] than those where bauxite is the starting point.

To make one tonne of aluminium, about 4½ tonnes of bauxite are refined, yielding 2 tonnes of aluminium oxide, this process requiring one tonne of fuel oil and 160 kg of sodium hydroxide. In the electrolytic process which follows, 450 kg of carbon anodes and 50 kg of cryolite are consumed, while the electrical power (a typical furnace requires 140 000 amps at 4.5 volts) amounts to 14 500–17 000 kWh/tonne of aluminium. The lower figure mentioned is obtained in the most efficient plants, where improvements having been achieved by more efficient electrode systems, greater capacity of pots and computer control of pot conditions.

The largest source of energy for aluminium smelting is hydroelectric power (54%); thermoelectric energy represents 44% and so far only 2% comes from nuclear energy. In individual countries the proportion may be vastly different from the world's average; for example Canada obtains all the energy from hydroelectric plants, while in Japan 87% is from thermoelectric power.

Until recently, aluminium smelters were able to purchase their electricity at favourable rates but now they are having difficulty in obtaining the same consideration. The ever increasing cost of coal, anticipated shortages of oil and ecological problems with nuclear energy, have added to the difficulties of primary producers who purchase their energy supplies. Those smelters who manufacture their own power are faced with the daunting much inflated capital cost of building new hydroelectric stations in mountainous areas.

Since most bauxite deposits are found in tropical or semitropical regions, remote from any industrial areas where the metal can be smelted, transport costs are a deterring factor. It has been calculated that the average distance travelled by aluminium oxide from ore deposit to smelter[4] is about 1 500 miles.

World wide research is continuing into the possibilities of extracting the metal from its ores with lower energy costs than are required at present. One process was announced by Alcoa in 1973 and a brief summary must fail to do justice to an effort which was the culmination of 15 years research work and the expenditure of about 25 million dollars. The first step[3] is the production of pure aluminium oxide by the Bayer process. This is then mixed with carbon and is chlorinated in a reactor at 700–900°C. The resulting gaseous aluminium chloride is passed through a special filter, sublimed in an inert gas and then converted to aluminium by electrolysis, while the chlorine is recirculated.

The corrosion problems caused by the chlorine and aluminium chloride have been immense and the conditions in the chemical reactors have been described as a plumber's nightmare. Although a plant has been set up at Palestine, Texas, it is surmised that the process will not become economically viable before 1983. Nevertheless a development such as this, which has the aim of producing aluminium with the expenditure of about 30% less energy than is required in the Hall-Héroult process, with a potential saving of 5000 kWh for every tonne of aluminium, is sufficiently challenging for strenuous efforts to continue.

Re-cycling of scrap

Practically all die cast aluminium alloys are produced from recycled materials. The amount of energy required is only about one twentieth of that expended in the electrolysis of aluminium oxide. Even the aluminium–silicon alloy LM6 (A.413.2), which has a limited amount of permitted impurities, can be made with a substantial proportion of carefully selected secondary metal. Primary metal is required only when this alloy is specified for aircraft components.

The scrap available to the secondary metal refiner can be divided broadly into two groups. Over half of the material consists of sheet off cuts from rolling mills, extrusion discards, and turnings. The remainder is obtained from manufactured products that have come to the end of their life and are available for recycling. Although in the long run the availability of secondary material depends on the production of primary aluminium, it is not valid to compare the amount of reclaimable old scrap with current production figures of primary metal, since the useful life of manufactured products varies widely. Milk bottle tops, coated laquered foil, toothpaste tubes and all-aluminium cans may be returned for recycling within weeks of their manufacture. Slags, drosses, and machined turnings from aluminium foundries become available within a few months. Parts of automobiles, aircraft or washing machines may not be available for 10 years or more. Aluminium overhead cables will probably not be a source of scrap for at least 50 years.

Most large consumers of sheet, extrusions and other primary material make contracts to sell discarded cuttings and other scrap of known composition to their suppliers, thus limiting the amount of high grade material available to the producers of aluminium alloy ingots for the foundry industry. If the supplies of new scrap start to decrease, the secondary industry has to pay more.

In some cases the reclaiming of aluminium scrap is made difficult because of the method of manufacture of the article. For example, in the UK there are many beer cans made partly of tinned steel sheet with easy-open tops of an aluminium alloy with 4.5% magnesium. Such a can contains steel, tin, aluminium and magnesium. In the USA a large proportion of cans are all-aluminium, and are more useful to the secondary metal refiner. Many Municipal Authorities in the USA encourage the inhabitants to bring cans and other scrap to well organized 'dumps' where the various materials can be deposited separately.

In Britain the all-aluminium can is a growing market. During 1980 about 20% of beer and soft drink cans were of aluminium, but by 1981 the proportion had risen to 41% for soft drink containers and 25% for beer cans. A typical aluminium can weighs only about 20 g, compared with twice that weight for tinned steel and as much as 230 g for a non-returnable bottle. Following American ideas one British aluminium manufacturer has begun a scheme for offering a small sum for returned aluminium cans. Those which have tinned steel bodies and aluminium–magnesium alloy tops are shredded, the ferrous metal separated magnetically and the aluminium alloy used for ingot manufacture.

Japan and several EEC countries which have thriving die casting industries require more secondary metal than those countries can provide. Their determination to obtain sufficient metal reduces the amount available elsewhere[5]. The scrap market is international and large buyers are going further afield to find additional sources. There is much financial speculation in scrap and some companies make more profit by the careful timing of their transactions than through the volume of their output[6].

Secondary aluminium alloy production

The production of aluminium alloy castings was well established at the beginning of this century, the most widely used alloy containing about 10% of zinc. British foundries made their own alloys from primary materials; however in Germany a secondary metal industry existed before 1920. The British secondary metal industry[7] began to grow in the 1930s, due to the efforts of refugees who came here. Metal refining techniques and larger and more efficient furnaces were developed to treat the increasing amounts of scrap as the use of aluminium and its alloys increased.

During the war it became essential to conserve the amounts of primary aluminium brought across the Atlantic; the secondary aluminium industry in the UK developed in two directions. The Ministry of Supply set up smelters to reclaim the metal from aircraft scrap, while commercial secondary metal refiners more than doubled their output from other sources. At the same time schemes for segregating different grades of scrap were introduced by the Government.

Soon after the end of the war refiners and founders jointly produced a set of specifications for a wide range of secondary aluminium alloys. Among these the alloy now known as LM2 was listed in about 1943, at first under an LAC specification, based on a composition easily produced from wartime scrap. Then a range of secondary alloys was approved, leading to British Standard 1490, which at first included 20 aluminium casting alloys. Ten of the original list have become obsolete, but others have taken their place, making a present total of 21 alloys, some of which are used for pressure die castings.

The handling and treatment of secondary metal has been improved continually[7]. Giant crushers and shredders, plants for cleaning swarf and processes for separating the aluminium which is usually entangled in drosses, have been developed. One of the most important stages in metallurgical control of secondary metal refining has been the use of equipment which makes it possible to analyse rapidly the molten alloys from the furnaces. A small sample is taken, and placed in a spectrographic analyser. Within less than a minute a print-out shows the amount of all the constituents in the metal so that adjustments to the composition can be made without delay. Most of the alloying elements, including silicon, copper and nickel, can be added direct into the bath of molten aluminium alloy, where they dissolve. Automatic ingotting and baling machines are now being used in secondary smelting plants while several manufacturers have built up services for delivering molten aluminium alloys to foundries, as will be discussed on page 51.

During the past 20 years productivity in the secondary aluminium industry has doubled but, in spite of this, the proportion of labour plus energy and other manufacturing costs has advanced. In 1970, about 80% of the cost of a secondary aluminium alloy ingot was from material but by 1980 this proportion had been reduced to about 75%, the manufacturing costs having risen to about 25%.

Secondary metal smelters are required to prevent the emission of noxious discharges. This requires supplementary equipment for extracting the furnace fumes and filtering them before the exhaust gases are allowed to pass into the atmosphere through a stack. The performance of the extraction equipment and filter bags is monitored electronically. The stack is also provided with photoelectric cells which check on the discharge, and signal if it contains substances liable to cause obscuration.

Aluminium scrap containing ferrous metals, for example castings with steel inserts, is melted and refined in sweating furnaces: molten aluminium is tapped from the furnace and the unmelted ferrous material is raked out. Machining turnings and swarf are crushed, dried and passed through a magnetizing section to remove ferrous metals, while foil and other thin pieces of pure aluminium are baled and usually melted in rotary furnaces of 5—15 tonne capacity. Scrap aluminium castings, forgings or other large pieces are melted and refined in reverberatory or electric furnaces.

The metal is treated with salt[8], usually sodium and potassium chlorides which, when mixed in equal proportions, form a eutectic composition melting at about $650°C$. Some smelters use natural salts, either those comprising equal amounts of sodium and potassium chlorides or those which contain about 2% calcium fluoride, the remainder being about two thirds NaCl and one third KCl. The amount of salt varies from about 10—40% of the weight of metal in the furnace, the proportion depending partly on the cleanliness of the material and partly on the efficiency of furnace operation. Impurities are absorbed in the salt, forming a slag which is run off. This contains some metal which should be reclaimed if possible. Nationwide efforts are being made to treat slags in order to separate metallic particles and to reclaim salt for further use. While suitable processes are being developed much of the slag has to be dumped.

Most aluminium foundry alloys have a relatively high silicon content, but part of the scrap used by the smelter is of nearly pure aluminium, so it is necessary to add silicon to the melt. The normal practice is to raise the temperature of the liquid metal to $800—850°C$, and charge part of the silicon in pieces about the size of a fist on to the surface of the melt, allowing it to remain there till it has reached metal temperature. It is then thoroughly stirred into the melt, either mechanically or using a nitrogen lance. The process is repeated several times until all the silicon has been added. Other elements such as copper and magnesium can be added as pure metal; when titanium is required it is added as a hardener alloy. Often it is necessary to remove magnesium from the alloy composition because wrought aluminium—magnesium alloy is available in greater amounts than the present requirements of the aluminium—magnesium casting alloys. There is a vast amount of technology centred on magnesium removal, usually involving chlorine but with strict control of furnace emissions to prevent the escape of the gas into the atmosphere.

A great deal of scrap comes from broken down machines; new processes have been developed to avoid wasting manual labour in taking the machines apart and sorting the various metals. The 'sink and float process' is intended for reclaiming the non-ferrous non-magnetic residues after the shredding the magnetizing process has removed iron and steel from the scrap. The Wemco system, operated successfully by several scrap processors in the USA, enables 90% of the aluminium, copper and zinc to be recovered. An article in *Materials Reclamation Weekly*[9], described the first British sink and float plant where automobile scrap is fragmented. After magnetic separation of the ferrous metals the material is flooded with water and passed over screens to remove dirt. Next, flock and other upholstery materials are removed in a separator where a fluid of specific gravity above 1.0 can be created by a rising current of water. Material lighter than the controlling specific gravity floats and overflows on to a dewatering surface. This 'float' material consists of plastics, glass, textiles, rubber, wood and any remaining dirt. Materials with a higher specific gravity, comprising principally non-ferrous metals, sink. After separation, the 'sink' particles are picked up by a rotating drum-type elevator which lifts and discharges them into the 'sink' side of the screen. The article referred to above compares the principle of heavy metal separation to the fact that cork floats on water while stone and steel sink, but by raising the specific gravity of the separating liquid to 3.0 the stone will then float, but the steel will sink. In early processes calcium chloride dissolved in water was used to raise the specific gravity but this has now been replaced in American plants by suspended magnetite and ferrosilicon in water to create a pseudo-liquid. Such mixtures can be adjusted to provide specific gravities ranging from 1.25 to 3.8. At the British plant, ferrosilicon is used to create a specific gravity of 2.7, at which density aluminium will float while stainless steel, copper alloys and zinc will sink. The remaining material is then hand picked, first removing rubber and other non-metallic substances that have not been separated in the first stage, after which stainless steel, zinc, brass and copper are hand-picked and put into separate containers.

In spite of claims by those outside the secondary metal industry that there are still plenty of untapped sources of scrap metal, it must be agreed that any such sources would involve more sophisticated refining methods and greater costs to make them available. With increasing difficulties in obtaining suitable scrap it becomes more costly to produce alloys with a minimum of impurities. This is a problem which is affecting the reclamation of all metals, leading to 'new looks' at specifications to ascertain whether they can be widened without deterioration of properties. The use of LM20 (A.413.0) in place of LM6 (A.413.2) with an increase of copper from 0.1 to 0.4% and modest increases in allowable iron and zinc, discussed on page 26 illustrate that a tight specification can be amended with safety, thus permitting the wider use of secondary materials. From the die caster's point of view the increased amount of iron reduces 'soldering' on the die and thus eases production, while the additional zinc gives slightly better machineability.

References

1. DOWSING, R.J.
   Aluminium second only to steel in worldwide engineering use. *Metals and Materials* p. 20 (January 1977)

2. TARGHETTA, D.J.
   Economic factors in aluminium production. *OEA annual conference, Madrid* (September 1970)

3. GRJOTHEIM, K.; KROHN, C.; MALINOVSKY, M.; MATIASOVSKY, K.; THONSTAD, J.
   Aluminium Electrolysis, pp.14−17, Aluminium−Verlag GmbH, Dusseldorf (1977)

4. DOWSING, R.J.
   Aluminium, from alumina powder to semi−finished product. *Metals and Materials* p.20 (February 1977)

5. (Editorial)
   Changing world of UK secondary aluminium. *Metal Bulletin Monthly,* No. 113, p. 27−31 (May 1980)

6. (Editorial)
   Secondary under study. *Metal Bulletin Monthly,* No. 108, p. 59 (December 1979)

7. GITTINS, M.G.
   The development of the UK secondary aluminium industry. *The British Foundryman, Special supplement,* p. 113 (June 1979)

8. SULLY, A.H.; HARDY, H.K.; HEAL, T.J.
   An investigation of thickening and metal entrapment in a light alloy melting flux. *Journal of the Institute of Metals,* **82,** 40 (1953−54)

9. (Editorial)
   Dunn Brothers installs new Wemco non−ferrous recovery equipment. *Materials Reclamation Weekly,* **137,** (24), 24 (February 21st 1981)

Chapter 4

Aluminium alloy die castings

Although silicon is soluble in molten aluminium, its solid solubility is very limited. The figures quoted in L.F. Mondolfo's book[1] are given below in *Table 4.1* but M. Hansen[2] reports even smaller solid solubilities at the lower temperatures.

TABLE 4.1

Temperature (°C)	Solid solubility (% by weight silicon)
577 (eutectic)	1.65
550	1.30
500	1.10
450	0.70
400	0.45
350	0.25
300	0.10
250	0.04
200	0.01
room temperature	less than 0.01

Figure 4.1 shows the equilibrium diagram of the useful range of aluminium–silicon alloys. The eutectic composition with 11.7% silicon is the figure for equilibrium conditions, but the composition can range from 11.7 to 12.7% silicon, depending on the cooling rate. A typical near-eutectic alloy, containing 12% silicon, cast and slowly cooled, consists of coarse flakes of silicon in an aluminium matrix. In this condition the alloy has low strength and ductility and, since the flow of metal during feeding is affected by clustered silicon particles, internal cavities tend to form. If the silicon particles become too large they present difficulties in machining by being torn away from the surface.

The effect of modification and rapid cooling

In 1911, J. Frilley[3] found that an aluminium–silicon alloy produced by the simultaneous electrolytic reduction of alumina and silica possessed a structure different from that shown when the two elements were conventionally alloyed. It is possible that the effect was due to the retention of sodium–aluminium fluoride (cryolite) from the electrolytic

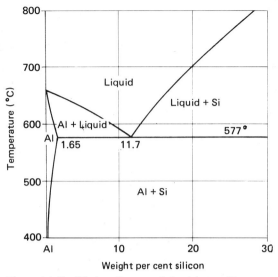

Figure 4.1 Equilibrium diagram of aluminium – silicon up to 30% silicon

reduction process. Ten years later Aladar Pacz[4] treated molten aluminium–silicon alloy with sodium and potassium fluorides before casting and found that the structure had been modified. In 1926, R.S.A. Archer and L.W. Kempf[5] reported that modification lowers the eutectic freezing temperature from 577°C to 564°C.

The addition of sodium, some sodium compounds, or other elements, including antimony and strontium, to aluminium–silicon alloys causes modification shown by a change from the coarse silicon flakes to a fibrous or rod form. Such an alloy has a substantially greater strength than the slowly cooled, unmodified alloy and the dispersed eutectic eases the problem of feeding.

In permanent mould and sand casting, the modification of aluminium–silicon alloys is necessary to obtain optimum properties. About 0.05% sodium is added, but only about 0.003% is retained. The treatment of the molten alloy is carried out either by metallic sodium, vacuum packaged in small aluminium containers to protect it from the atmosphere, or in the form of salts, generally a mixture of sodium chloride and fluoride with potassium chloride.

The modification and structure of these alloys is discussed by R. Elliott[6] in one of the International Metals Reviews and in a monograph in the present series of publications[7]. Some of the theories to explain the modifying effect of sodium and other metals and their compounds were also summarized by C.B. Kim and R.W. Heine,[8] based on their work at the University of Wisconsin. It had been observed that rapid cooling of the aluminium–silicon alloy produced a similar effect to that obtained by sodium modification and in 1957 R.C. Plumb and J.E. Lewis[9] examined modified structures with the electron microscope and concluded that the structures produced by rapid cooling and that by sodium modification are alike.

M.G. Day and A. Hellawell[10] studied the microstructure and crystallography of aluminium–silicon eutectics in terms of composition, freezing range and temperature. Sodium modification was reported to influence the growth mechanism of silicon but does not involve growth by a repeated nucleation process, as has often been suggested. It is envisaged that sodium restricts the growth of silicon by increasing undercooling, so permitting the silicon to grow at a lower temperature with a more regular, rod-like structure. The rapid cooling in pressure die casting also causes undercooling, encouraging silicon to grow in the same rod-like form. Cooling rate and sodium content work in the same direction and the resulting structure in a pressure die casting is similar to that formed in a slower cooled sodium-modified alloy.

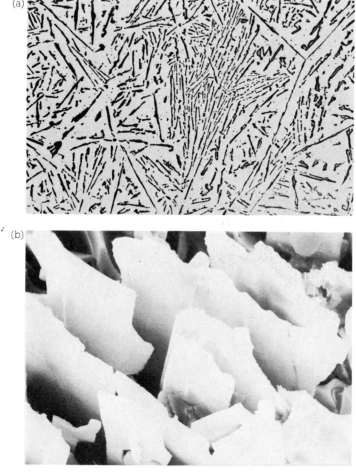

Figure 4.2 Unmodified aluminium – 12.7% silicon alloy. (a) Optical micrograph × 120 (b) Scanning electron micrograph of silicon plates after slow deep etching of the aluminium (× 1000)

Figure 4.2a shows an optical micrograph of a slowly cooled unmodified aluminium alloy with 12.7% silicon, while *Figure 4.2b* is a scanning electron micrograph of the same specimen after a slow deep etching of the aluminium to reveal the silicon flakes.

Figure 4.3a is an optical micrograph of the same alloy after modification and in *Figure 4.3b* the electron micrograph reveals the modified rod-like silicon structure.

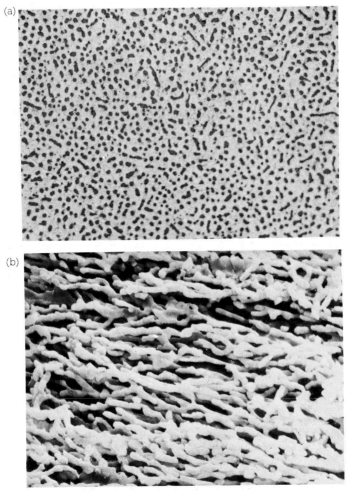

Figure 4.3 Modified aluminium – 12.7% silicon· alloy. (a) Optical micrograph ×120 (b) Scanning electron micrograph of rod-like silicon structure after slow deep etching of the aluminium (×500)

The aluminium–silicon alloys with about 12% silicon and a minimum of impurities display excellent resistance to corrosion and are recommended for such uses as food handling components and for castings which will be exposed to marine atmospheres. Such an alloy is specified as BS 1490 LM6 (A.413.2 in the USA). It has to be made from carefully selected materials, sometimes with the addition of primary aluminium. *Table 4.2* shows various specifications and permitted limits of impurity in aluminium–silicon alloys.

During the past 30 years a similar alloy, specified as LM20 in Britain, has been recommended and used as a very adequate alternative for LM6, allowing increases in the permitted contents of copper, magnesium, iron and zinc. In the USA the alloy is specified as A 413.0, in France as A-S 12 U. in West Germany as GD AlSi 12 (Cu), in Italy as

TABLE 4.2 Aluminium–silicon alloy designations

Country	Specification	Designation	Si	Cu(max)	Fe(max)
International	ISO DIS 3522	Al-Si 12	11.0–13.5	0.10	0.7
Austria	M 3429	G. Al Si	11.0–13.5	0.05	0.6
Belgium	NBN 436	SG-Al-Si 12	11.0–13.5	0.10	0.7
Canada	HA3	S 12 N	11.0–13.0	0.05	0.6
Denmark	DS 3002	4261	11.0–13.5	0.20	0.6
Finland	SFS 2566	G Al Si 12	11.0–13.5	0.10	0.55
France	NF A57-702 W	A - S 13	11.0–13.5	0.10	0.7
Germany	DIN 1725/2	G Al Si 12	11.0–13.5	0.03	0.3
Italy	UNI 4514	G Al Si 13	12.0–13.5	0.05	0.6
Japan	JIS H 2212	Dl V	11.0–13.0	0.05	0.6
Netherlands	NEN 6022	Al Si 12	11.0–13.5	0.05	0.5
Norway	NS 17510	Al-Si 12	11.0–13.5	0.20	0.6
Spain	UNE 38-203-76	L 2520	11.0–13.5	0.10	0.6
Sweden	SIS 144261	4261	11.0–13.5	0.20	0.6
Switzerland	VSM 10895	G Al Si 13Cu	12.5–13.5	0.08	0.7
UK	BS 1490	LM6	10.0–13.0	0.10	0.6
USA	ASTM B 179-80	A.413.2	11.0–13.0	0.10	0.6

GD-AlSi 13 Fe and in Switzerland as G-AlSi 13 Cu. The compositions of LM6 and LM20 are shown in *Table 4.3*.

TABLE 4.3

Alloy	Si	Cu	Mg	Fe	Zn	Mn	Ni	Sn	Pb	Ti
LM6(A413.2)	10.0-13.0	0.1	0.1	0.6	0.1	0.5	0.1	0.5	0.1	0.2
LM20(A413.0)	10.0-13.0	0.4	0.2	1.0	0.2	0.5	0.1	0.1	0.1	0.2

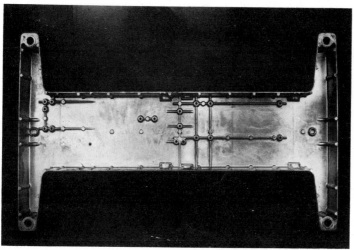

Figure 4.4 Sub-frame for a hospital bed. (Courtesy Horsel Werke and Joh Stiegelmeyer)

Figure 4.4 shows an enterprising use of aluminium alloy die casting which won the first prize for originality in the 1975 OEA competition. This sub-frame for a hospital bed, produced in the German equivalent of LM20, combined lightness with rigidity and was one of the largest die castings ever made, being 1260 x 125 mm. The wall thickness was 4.5 mm and the casting weighed 18 kg. It was delivered to the customer ready for installation and required only a primer and enamel coating.

The increase in allowable copper content of up to 0.4% in LM20 has been justified by a series of corrosion tests carried out in Germany and described on page 30. Both alloys have good resistance to corrosion but the purer LM6 is specified when corrosion conditions are severe; both have good strength and ductility. Although they are among the easiest aluminium alloys to cast by permanent mould and sand casting processes, they present minor problems to the pressure die caster in that they tend to 'solder' to the die more readily than the aluminium—silicon—copper alloys. This requires greater attention to die lubrication, more frequent die cleaning and, when possible, more draft on the diameter of small cores.

Aluminium—silicon—copper alloys

By far the greatest volume of alloys used by the pressure die casting industry is represented by the aluminium—silicon—copper alloys specified as BS 1490 LM2 and LM24 in the UK. The latter is equivalent to the USA specification 380.0 (A.380.1 ingots); there is no exact equivalent to LM2, but A384.1 is similar. LM24 has better machining properties than LM2 and its use has been increasing during the past decade; by 1980 it comprised about 22% of all the available aluminium casting alloys, compared with about 12% for LM2.

Automated processes for machining die castings, such as the components of automatic gear boxes, put a premium on speed of machining, so the trend towards LM24 could be expected. Nevertheless it is worth mentioning that LM2, containing a lower percentage of copper, has a slightly lower specific gravity than LM24. Thus if a 10 tonne batch of LM24 comprises 10 000 die castings, each weighing 1 kg, the same tonnage of LM2 would produce 10 182 castings, each weighing 0.98 kg.

It is good policy to select one of the two alloys, persuade customers accordingly and then to concentrate production on that alloy. If in the past both have been used, it is not difficult to transfer to the selected composition, though sometimes a slight runner alteration is required. Usually there is mild opposition from those who contend that better results were obtained from the eliminated alloy, but the concentration on one composition is more economic in many aspects, from metal purchase onwards. If a die casting company is obliged by customer reaction to continue using both alloys, it is desirable to stamp an identification on suitable positions of the die runner system, so that remelt materials can be separated.

Table 4.4 shows the compositions of the two alloys. The percentages from iron onwards are maxima. For comparison, the American specification for ingots of A.380.1, the alloy similar to LM24, is added. The American specification A380.0 for castings allows 1.3% for iron and 3.0% for zinc, representing slight increases above the ingot (A380.1) allowances.

TABLE 4.4

Alloy	Si	Cu	Mg	Fe	Zn	Mn	Ni	Sn	Pb	Ti
LM2	9.0-11.5	0.7-2.5	0.30	1.0	2.0	0.5	0.5	0.2	0.3	0.2
LM24	7.5-9.5	3.0-4.0	0.30	1.3	3.0	0.5	0.5	0.2	0.3	0.2
A380.0	7.5-9.5	3.0-4.0	0.10	1.0	2.9	0.5	0.5	0.35	Nil	Nil

Although eventually there may be some unification, there are at present many different designations and a fairly wide range in the permitted amounts of constituents. *Table 4.5* lists representative world wide specifications and designations for the die casting alloys comparable with LM24. To avoid too complex a chart, only silicon, copper, iron, zinc and manganese are shown. Permitted contents of the other elements vary widely; for example the Canadian alloy SC 84N is allowed only 0.05% magnesium, while SC 84R has 0.45–0.85%. In various other specifications nickel ranges from 0.03% to 0.5% while the content of lead ranges from zero to 0.3%, tin from zero to 0.2% and titanium from zero to 0.2%.

Many countries have different specifications for castings and ingots. For example the Japanese alloy specification JIS H 5302 designation ADC 10 is one of the alloys shown in *Table 4.5*. Their virgin ingots for the same material are covered by specification JIS 2212. D.10V, while JIS H 2118.D.10S is for secondary alloy ingots.

TABLE 4.5 Aluminium–silicon–copper alloy designations

Country	Specification	Designation	Cu	Si	Fe (max)	Zn (max)	Mn (max)
International	ISO DIS 3522	AlSi8Cu3 Fe	2.5–4.0	7.5–9.5	1.3	1.2	0.6
Belgium	NBN 436	DG AlSi8Cu3Fe	2.5–4.5	7.0–9.5	1.3	1.0	0.6
Canada	HA3	SC 84N	3.0–4.0	7.5–9.5	0.6	0.1	0.1
		SC 84R	3.0–4.0	7.5–9.5	1.2	1.2	0.5
Denmark	DS 3002	4254	2.0–4.0	7.5–10.0	1.1	3.0	0.5
Finland	SFS 568	G-AlSi9Cu3Fe	2.0–4.0	7.5–10.0	1.25	1.2	0.5
France	NF A 57-703	A-S9U3A-Y4	2.5–4.0	7.5–10.0	1.3	1.2	0.5
W. Germany	DIN 1725/2	G-AlSi8Cu3 (226)	2.0–3.5	7.5–9.5	0.8	1.2	0.5
		GD-AlSi8Cu3 (226D)	2.0–3.5	7.5–9.5	1.3	1.2	0.5
Italy	UNI 3601	G-AlSi8.5Cu	3.0–4.0	7.5–9.5	0.8	0.05	0.5
	UNI 5075	GD-AlSi8.5 Cu3.5Fe	3.0–4.0	8.0–9.5	1.1	1.0	0.5
Japan	JIS H 5302	ADC 10	2.0–4.0	7.5–9.5	1.3	1.0	0.5
		ADC 12	1.5–3.5	10.5–12.0	1.3	1.0	0.5
Netherlands	NEN 6022	AlSi8Cu3	2.5–4.5	7.0–9.5	0.7	1.0	0.6
Norway	NS 17530	Al-Si9Cu3	2.0–4.0	7.5–10.0	1.0	1.3	0.5
Spain	UNE 38-203-76	L-2630	2.5–4.0	7.5–10.0	1.0	3.0	0.5
Sweden	SIS 144251	4251	2.0–3.0	6.0–8.0	0.7	2.0	0.5
	SIS 144252	4252	2.0–4.0	7.5–10.0	1.1	1.2	0.5
Switzerland	VSM 10895	G-AlSi8Cu3	2.0–3.5	7.5–9.5	1.3	1.2	0.5
UK	BS 1490	LM 24	3.0–4.0	7.5–9.5	1.3	3.0	0.5
USA	ASTM B179-80	A380.0	3.0–4.0	7.5–9.5	2.0	3.0	0.5

The influence of copper in aluminium–silicon alloys

The aluminium–silicon–copper die casting alloys contain up to 4.5% copper; a solid solution of copper in aluminium is formed on solidification but, on cooling, the intermetallic compound $CuAl_2$ is precipitated from solution. The effect of copper on the corrosion resistance of aluminium–silicon alloys has been a matter of controversy but a series of tests carried out in Stuttgart[11] produced evidence that copper up to 0.5% is not harmful to the corrosion-resistance of the straight aluminium–silicon alloys. This research included the testing of 12% silicon alloys with copper contents from 0.01% to 1.0% in four different environments. In one test, water was evaporated and allowed to condense on the specimens, the heating was discontinued and the samples allowed to cool and dry; then the process was repeated once every 24 hours for a month. The second test involved a similar water condensation method but sulphur dioxide was introduced in the testing chamber. The third method was a salt spray test, and the fourth was carried out in a steam locomotive shed, giving a condition similar to a moderately aggressive industrial environment. The results indicated that samples with up to 0.24% copper showed resistance to corrosion equal to that of the copper-free alloy, while samples with as much as 0.52% copper were almost as resistant. However, the alloy samples with 1% copper corroded twice as much as those with low copper content. This emphasizes the fact that, while the aluminium–silicon–copper alloys with up to 4.5% copper are satisfactory for ordinary conditions of service, components which must endure severe corrosive conditions should be made in alloys with no more than 0.5% copper.

The effect of other elements

The specifications for the aluminium casting alloys developed during the past 40 years represent a sensible compromise between what is necessary to provide good mechanical properties, castability, and competitive prices, and the requirements of the secondary metal smelters who have to produce their alloys from available materials as economically as possible. The presence of some constituents decreases the resistance to corrosion because those elements are not solid soluble in aluminium, causing a second phase to be precipitated that affects the protective effect of the aluminium oxide layer. Other elements above permitted amounts lead to casting difficulties, such as segregation of intermetallic compounds. Fortunately, few of the elements that are likely to be introduced from secondary materials are deleterious. In that respect the makers and users of aluminium casting alloys are more fortunate than those concerned with zinc alloy die casting, where minute quantities of a number of impurities are positively harmful.

Iron up to about 1.3% is beneficial because it helps to limit the soldering effect of the die cast metal to the die[12]. Iron is attacked by molten aluminium and, although dies are protected by lubrication coatings, areas opposite the gate affected by turbulence of the injected metal become roughened and tend to adhere to the die cast metal. The presence of iron in the alloy reduces this tendency. However, the iron content should not be higher than the specified amount otherwise an intermetallic iron–aluminium compound, or complex intermetallics containing silicon, manganese and iron are formed which cause the type of problems which are discussed in Chapter 8.

A report by C.A. Queener and W.L. Mitchell[13] included a discussion of the high speed machining of die castings in the American 380 alloy, the equivalent of LM24. The work was undertaken to investigate problems in the machining of a die casting weighing

1½ kg on which an automatic machine performed 105 milling, drilling and tapping operations in 38 seconds. It was found that reducing the iron content from 1.4% to 0.7% made a substantial improvement in the life of the machining tools.

Since manganese is used extensively as a constituent of wrought aluminium alloys it is bound to enter into the composition of casting alloys made from secondary materials. Although this element has some solid solubility in aluminium, it occurs in the cast alloys as the intermetallic compound Al_6Mn. In a manganese-free alloy the A–Si–Fe constituent is in a finely divided form and does not lead to embrittlement. However, as manganese is usually present, intermetallic compounds are formed which, being heavier than the aluminium-rich alloy, sink to the bottom of the crucible and if entrapped in the die cast alloy, they cause difficulty in machining, as will be discussed on page 73.

The alloys LM24 and LM2 have allowable zinc contents up to 3% and the same amount is permitted in the American alloy 380-0 and in similar alloys in other countries. On the other hand some specifications call for a zinc content of the order of about 1% or less, as will be seen from the *Table 4.5* on page 29. Aluminium and zinc form a solid solution which has no pronounced effect on properties in the general purpose die casting alloys. In the low-copper aluminium–silicon alloys, zinc is limited to 0.1 or 0.2%.

The alloys which include magnesium as a main constituent are discussed on page 36. In the aluminium–silicon–copper alloys, the magnesium content is usually specified as below 0.3%, to limit the formation of the compound Mg_2Si. There is a deterioration in the tensile strength of aluminium–silicon–copper alloys when the amount of magnesium exceeds 0.3%.

An aluminium–nickel eutectic is formed at 6.4% nickel, melting at 640°C. It has a very small solid solubility in aluminium and is present as an Al–Ni compound or in complex constituents with manganese and silicon. In most pressure die casting alloys, nickel is treated as an impurity and is limited usually to under 0.5%. However, one special alloy, mentioned on page 37 contains nickel, manganese and titanium. There are some nickel–containing alloys suitable for permanent mould casting, but these require age hardening after casting.

In most aluminium casting alloys, chromium is restricted and is not listed in the specifications, but in LM28 and 29 up to 0.6% is included. Chromium is harmful in the aluminium–silicon copper alloys because it forms intermetallic compounds such as Al_7Cr which causes hard spots; the effect of this element on the tendency to form inclusions is discussed on page page 73.

Some casting alloys contain traces of titanium as a grain refiner, so secondary alloys may contain small amounts of the metal as an allowable impurity, though it is generally restricted to under 0.2%.

Die cast engine blocks

Cylinder blocks were produced in a pioneering effort by Doehler–Jarvis in the early 1950s. A paper by the late A.F. Bauer[14], given to the Society of Die Casting Engineers in 1960, outlined the stages of development at one of the Toledo plants. At first an experimental die was designed for a four cylinder engine intended for use by the Kaiser Corporation but it was then changed to six cylinder, the first die, weighing 25 tons was tested on a specially built 2000 ton machine in December 1955. Unfortunately the Kaiser project was discontinued but the experience gained was valuable when the first commercial production for the American Motors, Rambler was tooled in 1959 and put into production in 1960. Eventually four special machines were built and four dies made; the die castings were included in the American Motors' production for several years.

Dr Bauer's paper described some of the many problems which were overcome and the design changes that were necessary. The blocks were made in the 12% aluminium–silicon alloy; as many as 142 cored holes were included and over 70 water cooling channels were present in the die. Six centrifugally cast grey iron cylinder liners of the so called dry sleeve design were mechanically loaded into the die before casting. Considering that this project was initiated at a time when mechanization of die casting was in its infancy, the achievement was remarkable and, apart from die lubrication, many features of today's mechanization were applied.

Each shot required 70 pounds of aluminium alloy to be transferred into the shot sleeve by an air operated autopour. One operator had overall responsibility for the production of each machine, assisted by a man who placed the liners into the mechanized loader and helped to clean the die. A third man operated the clipping press and inspected

Figure 4.5 Renault engine with pressure die cast cylinder block cast on Buhler 2000 Parashot machine

the castings. The American Motors' die cast blocks with cast-in liner inserts were discontinued because the production and material costs at the time proved to be higher than those offered by cast iron, but the later Honda Civic four cylinder engine and Outboard Marine and Mercury Marine, both with two-stroke six cylinder engines have cast-in liners.

The Doehler—Jarvis effort prepared the ground for later developments. European blocks, in substantial production, are designed so that liners can be inserted after casting and lifted out for replacement or reworking; consequently that aspect of the production involves fewer problems than those experienced with cast-in liners. *Figure 4.5* shows a French cylinder block currently in production and *Figure 4.6* illustrates a cylinder block

Figure 4.6 Cylinder block produced as two half-castings. (Courtesy SIMI)

weighing 23.4 kg which was produced as two half-castings by an Italian company. This method of manufacture results in more simplified machining than is necessary with a one piece block. *Figures 4.7a* and *b* illustrate a Japanese motor cycle engine block.

(a)

(b)

Figure 4.7 A Japanese pressure die cast cylinder block for a 500 cc motorcycle engine. (a) The movable die. (b) The die casting being removed from the die. (Courtesy Honda Engineering Co Ltd.)

Hyper eutectic aluminium – silicon alloys

Aluminium–silicon alloys containing more than 12.7% silicon are classed as hyper-eutectic; on solidification they contain eutectic plus hard particles of primary silicon. For example the alloy with 20% silicon melts at 700°C, about 40°C higher than the melting point of pure aluminium and over 120°C higher than the eutectic. The freezing range is from 700°C to the eutectic temperature of 577°C.

When General Motors and Reynolds Metal began the development of the hyper eutectic alloy which culminated in the pressure die casting of the Vega engine block [15], they decided on an alloy designated as ASTM 390.1 (LM30 in the UK) with a composition 16–8% Si, 4–5% Cu, 0.6–1.0% Fe and 0.45–0.65% Mg. On account of the long freezing range of this alloy, General Motors applied their Acurad process [16]. This involves a double plunger method of injection so that an intensification of pressure allows additional metal to be fed during the alloy's freezing. The dies were provided with a large gate area, to ensure that the gate metal was the last to freeze. The die cavity fill rate, stated by General Motors to be between 0.5 and 4.5 seconds, was considerably slower than in conventional pressure die casting. The cooling of the die was arranged to ensure solidification from the furthest point of the die progressively to the gate.

While in conventional die casting pressures of 75–150 MPa (about 5–10 tons/in^2) are applied, the Acurad process employs injection pressures within the 40–120 MPa range. By use of the double piston, the application of enhanced pressure is withheld until the metal in the die begins to solidify; then the secondary piston comes into action, forcing additional molten metal in the large gate area into the die cavity, thus reducing internal shrinkage. Because of the slow fill time, the advantages of Acurad are limited to the production of die castings of thick section, such as the cylinder block.

The slow injection and comparatively slow production speed of this cylinder block made it essential to obtain correct size and distribution of the particles of primary silicon. Phosphorus was added to the alloy as an 8% phosphor–copper, which formed insoluble aluminium phosphide particles to serve as nuclei for the primary silicon. Additions of the phosphor–copper in sufficient quantity to provide 0.01% phosphorus in the alloy proved to be adequate. In order to obtain a finish on the cylinder block wall that would allow the hard silicon particles to stand proud, an electrochemical process was developed, firstly to remove a minute layer of aluminium above which the silicon particles could stand in relief, providing the hard wear-resisting surface on which the piston rings would ride. The piston skirts were given a sequence of treatments comprising a zincate layer, a copper layer and finally, iron plating.

Production of the Vega cylinder block was discontinued in 1976 owing to economic problems, but the development work that had been done must have opened the way for future production. A similar hyper eutectic alloy is being used for Porsche but their cylinder blocks are low pressure die cast. Outside the field of die casting, cylinder liners in hyper eutectic alloy are being produced by powder metallurgy using a process developed by Pechiney [17]. With this process the optimum size of primary silicon can be achieved, together with the addition to the matrix of solid lubricants such as graphite. These liners can then be inserted in engine blocks produced in conventional alloys.

Aluminium–magnesium alloys

The aluminium-rich aluminium–magnesium alloys are widely used in the field of wrought products and, to a more limited extent, in casting. These alloys have an attractive silvery white appearance, are strong and resist corrosion well. They can be anodized to form an oxide film that accepts a wider range of colours than the alloys containing silicon. The aluminium–magnesium equilibrium diagram is shown in *Figure 4.8* An alloy with 5% magnesium begins to freeze at a temperature about 70°C higher than that of the near-eutectic aluminium–silicon alloys and so the thermal fatigue of the die casting dies will be

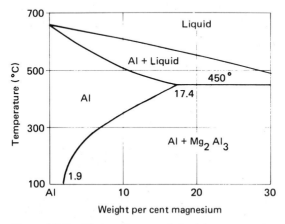

Figure 4.8 Equilibrium diagram of aluminium–magnesium alloys up to 30% magnesium

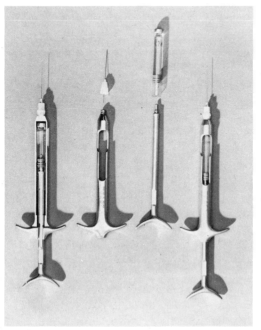

Figure 4.9 Dental syringe pressure die cast in aluminium alloy with 5% magnesium. The thread for the needle is cast. (Courtesy Anodic Castings Ltd.)

greater, leading to shorter die lives. If the 10% magnesium alloy is used, the melting point is a little lower than that of the 5% alloy but the freezing range is substantially longer. Consequently although this alloy can be sand cast and is not too difficult for permanent mould casting, it is not suitable for pressure die casting and all but the simplest shapes give difficulty with hot tearing.

Companies which have a long tradition of excellent production with the aluminium—silicon—copper alloys have difficulty in believing that the aluminium—magnesium alloys can be die cast with good quality, close limits of accuracy and satisfactory profitability. It appears that if die casters 'begin from scratch' with few preconceived ideas and concentrate their efforts on the alloy they intend to market, a higher level of success is attained than if a die caster introduces new alloys, with some trepidation, into his normal production pattern. *Figure 4.9* illustrates a dental syringe made of an assembly of pressure die castings in the alloy with 5% magnesium, produced by a British company which concentrates its production in the aluminium—magnesium alloys.

Other aluminium alloys

The published list[18] of British aluminium casting alloys ranges from LM0 for aluminium of 99.5% purity to LM30, a hyper eutectic alloy for engine blocks. In the course of time some of the series, for example LM3 and LM14, have been discontinued, but 21 designations remain in current use. Another list includes 17 aerospace alloys, cast by sand, permanent mould or low pressure die casting processes and usually heat treated. American, European and other specifications for die cast alloys are listed on various pages, including 27, 29 and 83. Some international organizations which circulate information on alloy specifications are mentioned on page 304.

From time to time other aluminium alloys have been developed; for example a material known as Pyral N2 contains 1.9—2.1% of each of nickel and manganese, with 0.1—0.6% of titanium and smaller amounts of allowable impurities. This alloy, which has a freezing range from 625—640°C, is specially suited for gas cooker burner heads; in spite of a casting temperature about 50°C higher than those of the standard aluminium casting alloys, Pyral N2 die casts sufficiently well for the small holes around the periphery of the burner head to be included.

Cast aluminium alloy wheels have been popular for several years, chiefly to improve the 'sporty' appearance or the performance of cars. Most of the wheels are produced as permanent mould or low pressure die castings but there is an acknowledged field for high pressure die castings, subject to the requirement of freedom from porosity. Either the wheels must be given stringent inspection to eliminate castings containing areas of microporosity or they must be produced by a process such as pore-free die casting which makes it possible to manufacture pressure die castings practically free from unsoundness.

A paper given in 1979 to the Society of Die Casting Engineers[19] outlined the possibilities of cast wheels and quoted one aluminium alloy, designated as 364 in the USA, containing 7.5—9.5% silicon; 1.50% iron; 0.2% copper; 0.1% manganese; 0.2—0.4% magnesium; 0.25—0.50% chromium; 0.15% zinc and 0.2—0.4% beryllium. This is similar to the Japanese alloy DX30 which also contains chromium, to alter the needle-like structure of the iron constituent and thus to improve ductility.

The French company Pechiney has developed a range of aluminium—silicon alloys to which 0.28—0.40% antimony is added. This causes a modification to the eutectic structure, making it fine grained, particularly when chill cast. Until 1981 these 'Calypso' alloys were used for permanent mould casting but further researches by Pechiney led to the development of Calypso alloy 49R specifically for pressure die casting. This contains 8.5—9.5% silicon, and 0.3—0.5% iron; copper is limited to under 0.02% and other impurities are similarly restricted. The alloy contains about 0.35% antimony and develops a fine grained structure, particularly under the rapid cooling conditions of pressure die casting; the same alloy can also be cast in the permanent mould process, but since its properties are developed by rapid cooling, it is not suitable for sand casting.

Calypso 49R has a freezing range 600—575°C. As with other die casting alloys the tensile strength depends on the thickness of the metal; for example a test piece 2.5 mm thick gives a tensile strength of 265 N/mm^2 while one of 6 mm thickness gives about 240 N/mm^2. The Brinell hardness ranges from 75 to 65 depending on the section of the test piece. It has a comparatively high elongation from 6—8% and has good resistance to corrosion. Like other aluminium—silicon alloys it can be anodized protectively, but the anodic coating is not suitable for tinted colour finishes for the same reason as for other silicon-rich alloys.

Castings in aluminium

During the mid-nineteenth century, a generation before Hall and Héroult had discovered the method of electrolytic extraction of aluminium from bauxite, numerous chemical processes were used for manufacturing aluminium in small quantities and at considerable cost. In a typical process aluminium was produced by mixing bauxite and sodium carbonate with sea salt and coal. The mixture was heated in a current of chlorine, which formed the double chloride of aluminium and sodium, this was then decomposed by sodium in a reverberatory furnace, using cryolite as a flux. At that time the possibilities of alloying aluminium with other elements had not been explored but numerous castings were made in practically pure aluminium, with a high standard of design and casting quality. The most famous of such castings is the statue of Eros in Piccadilly, cast in 1893, but long before that, other aluminium castings had been produced. The cap which rests on the top of the Washington Monument, cast in the 1870s, was made of aluminium with about 1.4% silicon.

An even earlier example was a copy of the statue of Diane de Gabies[20]. The original, about 2.25 m high, is in the Louvre but an aluminium model about 55 cm high was discovered in the 1950s by Dr E.G. West, then Director of the Aluminium Development Association. The plinth bore the name 'Paul Morin et Cie', a French company which produced many works of art, medallions, statuettes and jewellery at their factory near Nanterre. In 1860 the firm of Morin ceased to cast aluminium and turned their attention to copper alloys. This dates the Diane model to just before 1860. The aluminium casting had a composition including 3% copper, 1% silicon, 0.7% iron and small amounts of various impurities.

Today, apart from the pressure die casting of rotors discussed below, the casting of aluminium, either pure or with only small quantities of alloying constituents, is centered around the production of domestic utensils; these however are usually permanent mould cast or, in the case of kettle spouts, slush cast.

Pressure die cast rotors

The induction motor with squirrel cage rotor is one of the most important prime movers, both in domestic equipment and in industry, since it is versatile, robust, quiet and economical. Many millions of fractional horse power motors feature in washing machines, refrigerators, freezers, fans and electric sewing machines. A large and specialized industry is involved with the die casting of rotors in high conductivity aluminium for these small motors. Larger rotors[21, 22], sometimes up to 350 mm in diameter, are die cast for use in submersible pumps, lift motors and many other kinds of industrial drive motors. Generally motors up to 15 HP have pressure die cast rotors while larger ones may be either centrifugally cast, permanent mould or low pressure cast.

The rotor consists of a pack of closely toleranced slotted laminations from electrical sheet steel (typically 0.50 mm or 0.65 mm thick) and electrically conductive bars running through the slots. The combination of the bars joined at both ends to annular end rings is referred to as the squirrel cage. *Figure 4.10* shows examples of die cast rotors in which the iron laminations have been etched away to reveal the squirrel cages.

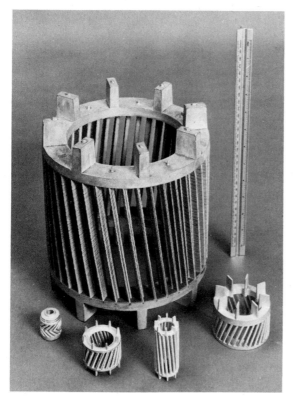

Figure 4.10 Diecast rotors; the laminations have been removed showing the aluminium squirrel cages. (Courtesy GKN Sankey Ltd.)

The laminations have a central hole for the rotor shaft and the pack is usually skewed about the shaft to give an angle with respect to the conductor bars. The pack of laminations is mounted on a removable mandrel or occasionally the motor shaft itself, skewing the pack to the required angle, enclosing the assembly in a die. Molten aluminium is injected so that it runs into the slots to form the conductor bars and into the cavities at either end to form the end rings, which may also incorporate cooling fins.

Usually aluminium of 99.6% purity is die cast, to obtain an optimum combination of conductivity and castability. The iron and silicon contents are specified as under 0.3% and 0.15%, respectively. This metal is prone to hot shortness, but that tendency can be lessened by further slight additions of iron, reducing the aluminium content to 99.5%. Since the molten aluminium has to be forced through lamination slots, the casting temperature may be specified as between about 720°C to 800°C, depending on the speed of ladling from furnace to die. The variation in metal temperature should not be greater than 20°C. The die temperature is held in the region of 250–300°C.

Producers of die cast rotors encounter typical problems which must be overcome if rejects are to be avoided. Cracks are caused if the metal is hot short or if the laminations are displaced. Shrinkage cavities may be caused during solidification, or by porosity from entrapped gas in the molten metal, in the dies or in the laminations. Partially cracked bars and small shrinkage cavities reduce the conductivity of the squirrel cage and hence the performance of the motor.

Small rotors for domestic appliances have varying numbers of laminations, depending on the size and power of the motor, but the weight of the lamination pack is always more than that of the die cast metal. Some plants are fully mechanized while in others one man is responsible for a sequence of operations. For example, the operator loads four lamination packs into the die, casts, removes the spray of four rotors, places the spray in an adjacent trim press, clips the spray and places the four rotors in a carton ready for delivery. An output rate of about 75 shots/hour, that is, 300 rotors, is obtained. In spite of casting

Figure 4.11 Diecasting machine with automobile lamination feed. (Courtesy Buhler Bros. Ltd.)

pure aluminium at about 750°C, die lives of about 50'000 shots are obtained. A steady routine of production is the key to satisfactory output and uniform quality. Producers of rotors have been somewhat remote from the mainstream of die casting but in the coming years they are bound to take advantage of the recent developments in technology and instrumentation.

Small high volume rotors are typically cast on horizontal cold chamber machines in the 250–400 tonne range. A Swiss machine with automatic feed of laminations is shown in *Figure 4.11*.

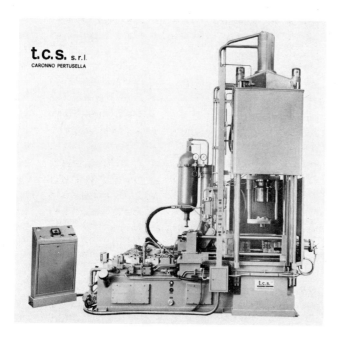

Figure 4.12 Vertical 50 tonne Rotorcast machine for producing rotors up to a diameter and length of 200mm. (Courtesy Alexander Cardew (Machines) Ltd. and TCS srl.)

For larger rotors, required in motors of over 1 HP, vertical machines are often employed such as the 'Rotorcast' illustrated in *Figure 4.12*. The aluminium is transferred to the machine by a pressurized metal transfer system and a Unimate robot carries out several operations, loading and unloading the machine, placing the cast rotors into a press for automatic clipping, then removing the cast rotors and placing them onto a pallet. Such a system produces rotors up to 132mm diameter.

In some vertical machines for manufacturing large rotors, as many as 1000 laminations are included, representing about 300kg of steel, on to which about 25kg of aluminium is die cast. The lamination pack is assembled and placed in the die cavity held above the metal container, the die is closed and the aluminium injected vertically upwards. Under such conditions the rate of output is comparatively slow; the heavy pack of laminations generally has to be loaded manually, but the molten metal can be transferred automatically and robots used for other stages in the process.

References

1. MONDOLFO, L.P.
   Aluminium alloys, structure and properties, 1st edition. (1976) Butterworths

2. HANSEN, M.
   Constitution of binary alloys, 2nd edition. (1965) McGraw–Hill, New York

3. FRILLEY, J.
   Researches on metallic silicon alloys and study of alloy densities. *Revue de Métallurgie,* **8,** 457 (1911)

4. PACZ, A.
   British patent No. 158 827 (1921) *US patent* No. 1 387 900 (1921)

5. ARCHER, R.S.; KEMPF, L.W.
   Transactions of the American Institute Mining and Metals Engineers, (73) p. 581 (1926)

6. ELLIOTT, R.
   Eutectic solidification. *International Metals Reviews* No. 219, pp. 161–186 (1977)

7. ELLIOTT, R.
   Eutectic solidification (in preparation)

8. KIM, C.B.; HEINE, R.W.
   Fundamentals of modification in the aluminium–silicon system. *Journal of the Institute of Metals,* **92,** 367 (1963–64)

9. PLUMB, R.C.; LEWIS, J.E.
   The modification of aluminium–silicon alloys by sodium. *Journal of the Institute of Metals,* **86,** 393 (1957–58)

10. DAY, M.G.; HELLAWELL, A.
    The structure of modified aluminium–silicon eutectic alloy *Journal of the Institute of Metals,* **95,** 377 (1967)

11. (communication from Vereinigung Deutscher Schmeltzhutten) translated by Association of Light Alloy Refiners, London, printed in *Metallurgia,* **77,** (461), 103–105 (March 1968) under the title Influence of copper on the corrosion resistance of Al–12% Si casting alloy.

12. GILLETT, B.A.; LATIMER, K.G.
    Influence of some alloying elements on tensile properties of pressure diecast aluminium test–bars. *Foundry Trade Journal,* **129,** (2821), 955 (December 31st, 1970)

13. QUEENER, C.A.; MITCHELL, W.L.
    Effect of iron content and sodium modification on the machinability of aluminum die castings. *Modern Castings,* **47,** 70 (February 1965)

14. BAUER, A.F.
    The first water–cooled die cast six–cylinder automobile engine block. *Society of Die Casting Engineers, First National Congress* Report No. 9C (November 1960)

15. JORSTAD, J.L.
    Development of the hypereutectic Al–Si die casting alloy used in the Vega engine block. *Society of Die Casting Engineers,* Congress Paper No. 105 (1970)

16. BRYANT, A.L.
    Acurad casting of hypereutectic aluminum engine block. *Society of Die Casting Engineers,* Congress paper No. 65 (1970)

17. PERROT, R.
Chemises de moteur en alliage d'aluminium hypersilicie, réalisées par métallurgie des poudres. (communication from Aluminium Pechiney, with summaries in English, German and Spanish.)

18. Association of Light Alloy Refiners, London. *The properties and characteristics of aluminium casting alloys.* (Technical publication, 146 pp.)

19. LEVY, S.A.; MILLER, J.C.; JORSTAD, J.L.
Techniques to improve the ductility of cast aluminum wheels. *Society of Die Casting Engineers,* Congress Paper No. G.T 79 − 074 (1979)

20. (Editorial)
The Diane de Gabies. *Revue de l'aluminium,* 198 p. 59 (April 1953)

21. MARKLEW, J.J.
Developments in die casting at the works of Brook Motors *Machinery and Production Engineering,* **115,** 886 (November 26th 1969)

22. HERRIDGE, F.W.
Die casting applied to large Newman induction motors *Machinery and Production Engineering,* **113,** 1036 (November 20th 1969)

Chapter 5

Aluminium as a substitute for cast iron

Aluminium has about one-third the density of iron, copper and zinc; this accounts for its wide use in transport and its anticipated future development. About 60% of all aluminium castings produced in Britain and the USA, 67% in Germany and Italy and 75% in France and Japan are for transport. Many of the remaining uses are for equipment where the light weight of the metal offers a significant advantage.

A decade ago large and heavy 'gas guzzlers' were normal in the USA but the Federal Government decreed that by 1985 the average fuel economy must be 27.5 m.p.g. The direct correlation between vehicle weight and fuel consumption[1] opened the way for the increased substitution of aluminium castings for heavier metals, particularly cast iron. In addition to the direct weight saving, the use of the lighter metal allows reduced power requirements and less weight in other components; for example a lighter car accommodates a lighter braking system.

In America, the Aluminum Association task force on automotive energy saving[2] surveyed the increased use of aluminium, indicating how the car of 1985 may be designed to achieve improved fuel economy. As two examples of articles written about the possible new uses of aluminium, the editor of Foundry[3] and D.E. Hatch and J.L. Jorstad[4] discussed inlet manifolds, cylinder heads, engine blocks, transmission cases and housings, brake drums, water pumps and power steering components, all of which depend at present on ferrous metals, especially cast iron.

European cars are smaller than the USA average, so the weight saving will not involve such a large increase in the use of aluminium, but the trend will be significant and is already providing opportunities for European die casters. The increase in tonnage of cast alloys will not be exclusive to pressure die casting; permanent mould and low pressure castings will have initial advantages since many of the parts to be replaced will previously have been cast iron, where undercut designs are not difficult to achieve. When such components are made in aluminium alloy, the permanent mould process can offer a similar undercut design, using sand cores, with very little alteration; low pressure die casting is also applicable.

The problems of converting a cast iron component to an aluminium alloy pressure die casting are illustrated by the efforts to produce inlet manifolds. The undercut portions of these manifolds in cast iron were previously made by sand cores and in most cases these curved passageways are difficult or impossible to form by metal cores. There have been exceptions; for example in 1962 the inlet manifold of the Hillman 'Imp' was geometrically designed with a complex series of core movements which enabled it to be produced as a pressure die casting, although at a rather low speed. In America a one-piece inlet manifold was produced as a die casting[5] for a Ford overhead camshaft engine.

For several years the Honda Engineering Company[6] has been producing small inlet manifolds by the use of expendable potassium chloride cores for making the curved passageways; after casting the core material is dissolved away. General Motors have experimented with a somewhat similar method, known as the lost foam process, described in a paper by D.R. Christman,[5] but that process appeared to be suitable for only sand, permanent mould and low pressure casting.

Another approach to the production of inlet manifolds is by welding together two pressure die castings. The six cylinder Chrysler intake manifold[7], designed in 1960, was initially cast in grey iron but in the interest of fuel economy, and to take advantage of the high productivity of die casting, it was decided to make the manifold by joining two pressure die cast components. Steel wire staples and sealants were evaluated as possible methods of joining the die cast halves but the manifolds were unable to withstand the stresses when machined on the existing tools that had been satisfactory for the cast iron manifolds. During the late 1970s a method of welding was developed to join the two halves, using a computer-controlled electric beam welder. One key factor in the success of this operation was the use of helium effluent instead of the usual compressed air. The two parts of the manifold were designed with a high single plane split line and a nested weld flange. The top casting had a flat surface to seat on a horizontal shoulder in the bottom casting; both pieces had 10 degree slanted walls which were self registering when pressed together prior to welding. The floor to floor time for assembly was about 26 seconds. Each die cast assembly represented a weight saving of 7 kg over its cast iron counterpart and a corresponding 30% cost reduction.

Future mass production of die cast inlet manifolds will be stimulated if the castings are made free from microporosity, which causes the risk of weld seam leakage. The paper by D.R. Cristman referred to above suggests that the pore-free process, which will be discussed on page 235 permit die cast multi-piece components such as inlet manifolds to be joined by welding without the danger of leakage.

In spite of the savings in fuel to be achieved by reducing the weight of cars, the amount of energy required to produce aluminium versus iron, must be taken into consideration and, in the short term, iron has the advantage. When the energy expenditure is added up, from quarrying the ore to a delivery of pig iron ingots, 1 kg of cast iron requires the equivalent of only about 5.3 kWh of energy. A kilogram of aluminium from ore to ingot requires over 60 kWh and, even taking into account the possible savings discussed on page 18 it is not likely that the total quary-to-ingot energy will fall much below 50 kWh/kg. In comparing the costs of different materials the volume of a component is more relevant than its weight; nevertheless the total amount of energy to produce a unit volume of primary aluminium will be nearly six times greater than that required for cast iron. Admittedly, the die casting industry uses secondary metal, which requires only about one-twentieth the amount of energy involved in producing the primary metal, but reclaimable aluminium scrap originally came from primary aluminium. However, the energy saving of ferrous metal smelting is a 'once only' advantage, while the saving through using the lighter metal is continuous. This was illustrated in a paper by Kurt Honsel given to the 1979 OEA General Meeting in Madrid[8]. 3000kg of aluminium replaced 6000kg of steel in the Brussels underground system. The power saved per annum amounted to 18 000 kWh, and although, initially, less energy was required to produce the steel, the annual energy saving would move in favour of aluminium and, since the life of the system is expected to be over 30 years, the accumulated benefit of the change to aluminium will be substantial.

A great deal of evidence points to a substantial increase in the use of aluminium and the production of die cast aluminium alloys. Unfortunately although the potential for growth is proven, the achievement of this growth will require something more than the enthusiasm and ingenuity of die casters. As we have discussed in an earlier chapter, the building of additional smelting capacity, linked with comparatively economic sources of energy, involves an enormous expenditure of capital and it is possible that aluminium smelters will wait until they are confident of substantially increased markets before taking steps to provide the required greatly increased amounts of aluminium. The secondary metal market, on which die casters depend, is eventually affected by the amount of primary metal produced, so periods of shortage and consequent raw material price increases can be expected.

References

1.   PALAZZO, F.
      The future of aluminium in the automotive industry. *Organisation of European Aluminium Smelters General meeting*, Florence, p. 17 (23rd September 1977). Association of Light Alloy Refiners Ltd., London.

2.   The Aluminum Association (USA) Task Force on automotive energy saving *Use of aluminum in automobiles – effect on the energy dilemma*. 3rd edition. (April 1980). Washington

3.   MISKE, J.C.
      Carmakers turn to aluminum castings. *Foundry Management and Technology*, 106, 26 (February 1978)

4.   HATCH, D.E.; JORSTAD, J.L.
      Aluminum structural castings result in automobile weight reduction. *Society of Automotive Engineers Congress*. Detroit, (February/March 1978) Paper 780248

5.   CHRISTMAN, D.R.
      Casting process selection for light metal parts. *Society of Die Casting Engineers* Congress Paper No. G–T77–083 (1977)

6.   NAKASHIMA, W.; EBISAWA, M.
      Die casting of major engine components. *Institution of Mechanical Engineers*, 194, (5) (1980) (Automobile Division)

7.   LEE, J.; GOLIN, L.
      The Chrysler die cast and welded aluminium 6–cylinder intake manifold. *Society of Die Casting Engineers* Congress Paper No. G–T79–082 (1979)

8.   HONSEL, K.
      Development and future prospects of the aluminium foundry industry. *Organisation of European Aluminium Smelters General meeting*. Madrid (September 1979). Association of Light Alloy Refiners Ltd., London

Chapter 6

Silicon

At an early stage in the development of the aluminium casting industry it was discovered that silicon, particularly within the range 6–15%, forms useful alloys with aluminium. A historical survey by E. Scheuer[1] traced the development of the aluminium–silicon alloys back to the 1850s. Although silicon plays such an important part in aluminium casting, its properties and processes of manufacture are taken for granted to a greater extent than most other alloying elements.

Silicon is a hard and brittle semiconductor element with an atomic lattice structure similar to that of carbon in its diamond allotropic form. After oxygen, silicon is the next most plentiful element in the earth's crust; it is a constituent of sand, quartz, clays and most rocks, but the elemental silicon does not appear in Nature. Its remarkable electrical properties, both in the pure, single crystalline state and in the presence of minute amounts of added elements such as boron, have led to the present revolutions in microprocessing. Its tendency to form complex molecules, as for example the silicones, has similarities to the chemical behaviour of carbon. It has a melting point of 1410°C. In ferrous metallurgy silicon is usually added as ferrosilicon, but pure silicon is added when making up aluminium–silicon alloys.

Silicon is usually manufactured in mountainous areas where ample supplies of hydroelectric power are available. The growing markets for the element have led to its bulk production in electric arc furnaces, many of them operating at over 20 000 kVA. The process of reduction is theoretically simple, depending on the endothermic reduction of silica by carbon.

$$SiO_2 + 2C = Si + 2CO$$

However in practice the process is complicated by secondary reactions, including the following.

$$SiO_2 + 3C = SiC + 2CO$$
$$SiO_2 + C = SiO + CO$$
$$2SiO_2 + 4C = SiO + SiC + 3CO$$
$$SiO + 2C = SiC + CO$$
$$2SiC + SiO_2 = 3Si + 2CO$$

The essential raw material is quartz which must be of high grade, preferably with less than 0.5% total impurity. The reducing agents must not only provide the carbon but must give the required conductivity to ensure the passage of electricity in the furnace and to form a charge which is sufficiently porous to facilitate the reactions. The carbon is provided in several forms: bituminous coal, petroleum coke, wood chippings (preferably from beech, chestnut or birch trees), and sometimes charcoal is added. The proportions of these materials vary according to the manufacturing routines of the various plants, but the total weight of reducing agents is always about the same as that of the quartz. *Table 6.1* shows the amounts used in three typical reducing plants, expressed as kg/tonne of quartz.

TABLE 6.1

	1.	*2.*	*3.*
Petroleum coke	100	300	300
Bituminous coal	450	200	50 (charcoal)
Wood chippings	450	500	750

The raw materials are obtained from many parts of the world; for example a Norwegian plant near Trondheim imports quartz mainly from Spain, coal, petroleum coke and prebaked carbon electrodes from the USA, and wood chippings locally from Norway. The required electric power is generated from the Company's own hydroelectric plants. Another large producer, in France[2], obtains the quartz, charcoal and wood chippings from local sources, while prebaked electrode requirements are covered from France, Italy and USA. Petroleum coke is imported. Other producers of silicon include Spain, Brazil, Portugal, Italy, Switzerland, Sweden, India, Japan, USSR and Yugoslavia. In future Australia may also become a major producer.

Smelting is carried out in large submerged rotation electric arc furnaces with steel shells lined with refractory bricks and carbon blocks. The rotation of the furnace counteracts the tendency for a restricted reaction zone to develop from silicon carbide build ups. The massive electrodes dip into the mixture of quartz and reducing agents. One of the largest silicon smelting furnaces in the world, at the Norwegian plant, operates at 27000 kVA with a voltage of 250 and an amperage of 85000. In this furnace the electrodes are 1.45 m in diameter. The furnace produces about 45 tonnes of silicon per day with a recovery level of 70–80%. Depending on the operating conditions, silicon reducing furnaces operate with a power consumption of between 12000 to 14000 kWh/tonne of silicon[3]. Between 90 to 100 kg of carbon electrode is consumed per tonne of silicon.

One of the many problems encountered in the production of silicon arises from the high vapour pressure of silicon monoxide which is formed according to two of the formulae on page 47. The gas is partly lost to the atmosphere, causing the typical grey smoke from silicon and ferro-silicon furnaces. The condensation of silicon monoxide causes a sticky layer in the upper part of the furnace which temporarily prevents gases from escaping. Finally the gas will burst through and silicon is lost as amorphous silicon. To counteract this, the sticky layer has to be broken up at regular intervals.

The molten silicon is tapped from the furnace and after solidification it is crushed in jaw crushers or pulverized in impact mills. At this stage the crushed or powdered element is about 85% pure. A leaching treatment is given to purify the silicon; a typical analysis is as follows:

Silicon	98.3%
Iron	0.5%
Calcium	0.25%
Aluminium	0.75%

The remainder of the impurities comprise small amounts of manganese, titanium and oxygen.

References

1.   SCHEUER, E.
      History and development of aluminium silicon alloys. *Foundry Trade Journal*, p.2—11 (May 31st 1951)

2.   (Editorial)
      L'elaboration due silicium metal au four électrique. *Journal du Four Électrique*, pp. 225—227 (November 1971)

3.   FAIRCHILD, W.T.
      Production of silicon metal in the electric furnace. *Journal of Metals,* **22,** (8), 55—58 (August 1970)

Chapter 7

Melting equipment for aluminium alloys

Technical progress is characteristic of any virile industry but it is necessary to reassess the effect of economic factors regularly. The design and selection of metal melting equipment for die casting exemplify how changing conditions have stimulated progress. Three factors have emphasized the need for improved melting equipment: mounting costs of energy, inflation, which caused increased raw material prices, and the development of automation, which necessitated design changes in plant layout.

In order to set a target for the efficient use of energy in melting aluminium alloys, the theoretical energy consumption can be calculated. The specific heat of aluminium ranges from 0.215 at 20°C to 0.286 at 650°C. An average figure of 0.25 has been taken for the purposes of the calculation. The latent heat is 93 cal/g. The amount of heat required to bring one gram of aluminium–silicon–copper alloy to casting temperature can be divided into the following stages:

1. Heating the metal from room temperature (10°C) to melting point (580°C)
   mass (1 g) x specific heat (0.25) x temperature difference (570°C)   *142 cal*
2. Converting the solid metal at melting point to liquid
   mass (1 g) x latent heat (93 cal/g)   *93 cal*
3. Heating the metal from melting point (580°C) to casting temperature (650°C)
   mass (1 g) x specific heat (0.25) x temperature increase (70°C)   *18 cal*

This makes a total of 253 cal/g, or 253 000 kcal/tonne.

The heat capacity of fuel oil is measured in BTUs, electric power in kWh and gas in therms. The following conversions can be used:

1. One litre of gas oil provides 36 400 BTU. One BTU is equivalent to 0.252 kcal, so one litre of oil provides 9173 kcal.
2. One kWh of electricity provides 3412 BTU, equivalent to 860 kcals.
3. One therm of natural gas provides 100 000 BTU and therefore one therm has a heat capacity of 25 200 kcal.

Thus ideally it would require 27 litres of oil, 294 kWh or 10 therms to bring 1 tonne of aluminium alloy to casting temperature.

The cost of fuel oil has multiplied 10 times during the past decade; natural gas appeared likely to become a powerful competitor against oil for melting but considerable price increases, coupled with restriction of supplies in the UK, made gas less favourable than had been anticipated. Electric melting, once thought of as being limited to situations requiring specially high quality control, now competes on a cost basis and is favoured for ecological reasons.

As the cost of energy escalated, die casting companies realized that melting furnace efficiencies of 10–15% could not be tolerated. Therefore, more efficient furnaces, better burner designs, heat recuperation and improved refractories have helped to counteract soaring fuel costs. Having achieved efficient furnace design it is always desirable to have continuous production, for when a furnace is idling it is using energy, although at a reduced rate. The conditions of short time working during industrial depressions emphasize the lowering of efficiency that follows.

In the years when metal costs were low, the loss of a few per cent by drossing or other forms of metal wastage may not have appeared serious, but in recent years foundry managements have been stimulated firstly to identify metal losses and then to explore methods of reducing them. The measurement of the metal loss in a single melt is easy enough; the calculation of total metal losses per annum, discussed on page 189 is more difficult but more realistic. Furnaces that offer a saving in melting losses may now be competitive, even though their initial cost is greater than that of less efficient equipment.

The past decade has seen great advances in mechanized production. This had already begun in the 1950s with zinc alloy, but world wide progress towards automatic aluminium die casting was taking place during the 1970s and still more advances are inevitable in the 1980s. The efforts to automate aluminium die casting received an added stimulus through weight saving campaigns in the automobile industry. Since that involved aluminium competing with cast iron for the production of massive components, mechanized production was necessary. The cold chamber machine must work in conjunction with automatic molten metal feed, a sequence of die lubrication and a robot or other form of handling device to remove the casting from the die cavity. Consequently the size, shape and capacity of the holding furnace must be tailored to become part of the automatic system.

The provision of metal takes place in five stages: delivery of ingot or molten metal from the smelter to the die casting plant; bulk metal melting at some central area; transport of molten metal to the die casting machine; holding of molten metal at each machine, and the return of runners and scrap to the bulk melting section.

Molten metal deliveries

The amount of heat required to melt 1 tonne of aluminium alloy and bring it to casting temperature will certainly be more than the theoretical 253 000 kcal. That heat must be used over again when the die caster has to remelt metal previously cast into ingots at the smelting plant. Under suitable conditions, molten aluminium alloy can be delivered, thus avoiding the remelting of ingots and saving energy, labour and melting space. Molten metal is delivered[1] either in ladles which are emptied on arrival at the die casting foundry, or in a dispensing unit which is left, while an 'empty' is taken back to the smelter. This second method is preferred in most cases. *Figure 7.1* shows typical transport ladles of about 3 tonnes capacity, the size mostly used in the UK. In the USA ladles holding about 5 tonnes are customary. The tare weight of the empty ladle is about the same as the metal capacity and the lid weight is about 1 tonne, so that for a 3 tonne load of molten metal the total weight is about 7 tonnes. Transport is by trailers which carry two or three ladles.

Figure 7.1 Transport ladles for aluminium alloys. (Courtesy Alcan International Ltd.)

The temperature drop during transport is 20–30°C/h so, to avoid too great a superheat being necessary, a journey of 5 h should be the limit. Ideally, plants which receive molten metal should be close to the smelter but deliveries in Britain can be made up to 100 miles away. A recent American article[2] suggested that the distance should be limited to 300 miles, but the statement was qualified by the observation that the metal would then have to be superheated by 200°F (about 100°C). Such long journeys may indeed be possible but the quality of metal delivered is bound to suffer. The transport of molten aluminium alloy is covered by safety regulations, including the identification of the vehicles and training of drivers in the routine for delays and emergencies. To make molten metal deliveries viable, it is necessary that bulk supplies of the same alloy are required. Special capital equipment is needed at the die casting plant and a price premium is charged by the supplier to cover his equipment and transport.

In the UK the proportion of molten metal versus ingot deliveries has reached a more or less static situation (about 10% of the total tonnage). Companies which require large amounts of aluminium alloy and which are suitable for molten metal deliveries from a geographical point of view continue to take all or part of their requirements of molten metal, while the rest take ingots. In the USA, about a third of the aluminium needed by die casting companies comes in molten form.

Bulk melting

Companies that purchase their aluminium alloy as ingot and recycle their own process returns select an area near the foundry where the metal is bulk melted and then distributed to the individual die casting stations. Companies which take molten metal deliveries need to have additional equipment for remelting scrap and runners, although a few manufacturers return such material to secondary metal refiners for melting, cleaning, chemical analysis and re-delivery. There is an agreed allowance for metal loss during this treatment, depending on the circumstances.

The several types of bulk metal melting furnaces suitable for die casting plants can be divided into two groups: furnaces which are used continuously for long periods, melting only one alloy, and those which are designed for batch melting, the furnace being emptied after each charge. Small die casting companies, particularly those which handle several aluminium alloys, may use basin tilting furnaces for batch melting. At the other extreme the large producer with continuous shift working may prefer reverberatory furnaces which, in spite of comparatively high metal losses, have the merit of rapidly melting large amounts of metal.

During the past decade other furnaces have been developed as bulk melters. Electric induction furnaces offer the advantage of low metal losses, efficient use of energy and rapid throughput, though with a comparatively high capital cost. A recently developed method of melting, the immersed crucible furnace, can be used either for bulk melting or as a melter holder at the die casting machine.

Reverberatory furnaces

Several types are available; some are capable of receiving molten metal deliveries or for melting large tonnages of ingot and bulky scrap. Others have an outside well suitable for receiving a variety of scrap, runners and ingots. *Figure 7.2* shows an American furnace of 10 tonnes capacity. A British reverberatory furnace with a similar capacity and a melting rate of 4 tonnes/h is shown in *Figure 7.3*. The materials to be melted are charged through a large end door of the vertical lift type. Firing is by natural gas; fuel efficiency is aided by a heat exchanger with stainless steel tubes set in a refractory lined chamber, in which the hot exhaust gases preheat the combustion air up to about 400°C. This furnace is tilted for pouring molten aluminium alloy by actuating the single-acting hydraulic cylinders fitted to the rear of the furnace body.

Plants making large castings in considerable quantity require furnaces with capacities of 30 tonnes or more. In America, such reverberatory furnaces are often built up on site. A suitable design is selected and the steel shell is ordered, cut to size, shipped to the die casting manufacturer and welded and assembled on the floor. Refractory bricks are purchased from another supplier who provides the necessary expert labour to brick the furnace. Furnaces built up in this 'do it yourself' manner cost about 25% less than furnaces purchased ready built.

Another type of reverberatory furnace, illustrated in *Figure 7.4,* provides improved access to the lining for furnace cleaning and maintenance. The furnace is cylindrical and has a large access door at one end directly under the flue for fluxing and cleaning. Cold air entering the furnace when the door is opened passes upwards through the flue and so

Figure 7.2 A 10-tonne capacity reverberatory furnace. (Courtesy M.P.H. Industries Inc.)

Figure 7.3 Monometer reverberatory furnace

Figure 7.4 Reverberatory furnace with large access door for cleaning.
(Courtesy Morganite Thermal Designs Ltd.)

does not come into contact with the surface of the metal. The cylindrical body is lined with standard refractory bricks, thus overcoming one of the difficulties of relining reverberatory furnaces, where bricks often need extensive cutting before they are put in position.

Dual energy reverberatory furnace

In a paper given to the Society of Die Casting Engineers congress in 1969, Scott. P. Kennedy[3] described a furnace which can use electricity, gas or oil. Its development was based on the successful operation of radiant roof furnaces, described later. When operating by electricity the furnace used an upper chassis with electric radiant elements, and the flue and burner ports were plugged. If it was decided to convert to gas or oil, the electric panel was shut off and the heater elements removed. Fibre plugs were installed in the element holes, the insulator burner plugs were removed and the burner nozzles connected. The furnace could now operate with natural gas, propane or oil. In converting back to electric power, the process was reversed, the changeover taking 1–4 hours.

Immersed crucible bulk melting furnace

This type of furnace, developed in Britain is suitable for bulk melting, but a much more important use is at the individual die casting machine stations, discussed on page 65. It is a new furnace concept: the burner fires into a crucible immersed in the molten metal contained in a refractory bath as illustrated in *Figure 7.5* with a schematic section also shown. The crucible is identical to those used in bale-out furnaces; it does not hold metal but is surrounded by metal. A high intensity flame makes it possible to complete combustion within the confines of the crucible. The exhaust gases are used to preheat the roof over the melting end of the bath of metal and the solid charge of metal in the flue. These furnaces can be charged mechanically without damage to the crucible.

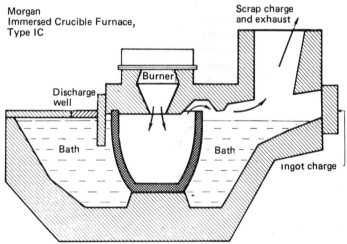

Figure 7.5 Immersed crucible furnace. (Courtesy Morganite Thermal Designs Ltd.)

Melting capacities of 250–850 kg/hour are available. The furnace is charged through the flue with ingot or scrap. When the metal in the lower part of the flue is melting, it can be pushed into the bath with a suitable rake through the rear access door. For the most efficient operation the flue should never be allowed to empty during the melting process. When melting at maximum rate the efficiency is in excess of 40%.

To achieve the desired melt rate and best life from the crucible and refractories the flue should be kept filled with metal during melting and the furnace should be cleaned thoroughly at about 4 hour intervals[4]. Although this furnace is versatile and efficient the regular programme of cleaning must be carried out, but in well-organized plants this should take only a few minutes. When one is provided with equipment that gives fast and thermally efficient melting it is all too easy to assume that it can be left to look after itself. There have been instances where managements have neglected good housekeeping rules in mechanized die casting foundries using immersed crucible furnaces with a need-less loss of efficiency. On the other hand, those who have operated the furnace with attention to careful maintenance have obtained excellent results.

Radiant roof furnaces

Using radiant heat from the roof and walls, radiant roof furnaces are suitable for melting foundry returns or to receive molten metal direct from secondary metal refiners. They utilize flat flame burners in the roof, which do not disturb the oxide film on the surface of the metal. They have a capacity to melt ratio of 10:1; thus a 9000 kg furnace would melt at about 900 kg/hour. Melt losses are lower than for convential reverberatory furnaces. *Figure 7.6* shows an American radiant roof reverberatory furnace.

Basin tilting crucible furnaces

Small and medium sized producers, especially those working single shift, or which cast a range of aluminium alloys, find that basin tilting furnaces are ideal for their bulk melting. A typical furnace, capable of melting about 500 kg/hour, is illustrated in *Figure 7.7*. Ingots and scrap can be charged into the crucible and, when the metal is molten, the furnace is tilted to pour metal into a portable ladle, which is conveyed to the die casting machines by monorail, mobile ladle or fork lift truck. Melt losses are appreciably lower

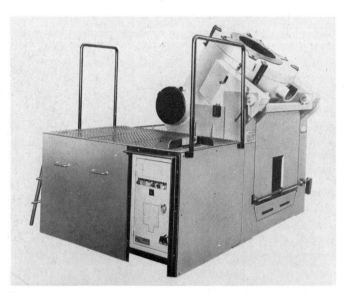

Figure 7.7 Basin tilting crucible furnace. (Courtesy Morganite Thermal Designs Ltd.)

Figure 7.6 Radiant roof furnace. (Courtesy M.P.H. Industries Inc., and D.K. Furnaces Ltd.)

than those obtained with reverberatory furnaces. The problem of mechanized charging of ingot and scrap has not been solved, because it introduces the hazard of crucible damage. Basin tilting furnaces obtain increased efficiency by supercharge preheating. The products of combustion are taken over the top edge of the basin and pass through the solid charge protruding above the crucible. The flame is not in contact with the molten metal, so losses are kept to a minimum. Electric resistance basin tilting furnaces are available, heated either by wire elements or silicon carbide rod elements. Melt times, even with silicon carbide elements, are longer than with fuel fired units, largely because there is no supercharge preheating.

The basin tilting furnace provides excellent thermal and metallurgical efficiency, but it is not suitable for the large scale producer who places emphasis on mechanization. In such cases reverberatory furnaces are often preferred; however, more and more die casting companies are finding that electric melting provides better quality and lower melting losses than reverberatory furnaces, though at a higher capital cost.

Electric induction furnaces

Although electric induction furnaces[5] have been used for melting iron and copper alloys for many years, their use in aluminium alloy die casting was limited, partly because the operating cost was higher than that of oil and partly because various technical problems had not been overcome. However, in the last 10 years, improvements due to the introduction of solid state controls, simpler circuitry and fewer components, have led to a reduction in manufacturing costs, opening the way for the extended use of induction furnaces for the bulk melting of aluminium.

Induction furnaces provide molten metal of higher quality than that melted in most oil or gas fired furnaces; also the metal losses are lower. In contrast to electric resistance or fossil fuel fired furnaces, in which heat is transferred by radiation, the power in the furnace is induced into the melt. The range of frequency is from the normal 50 Hz for very large melting units, through medium frequency of 1000 Hz for most die casting bulk melting, to high frequency of 10 000 Hz for melting small amounts of metal in the laboratory.

There are three types of induction furnaces suitable for aluminium melting — channel furnaces, coreless furnaces and channel-less detachable inductor furnaces, the last being a new development. The channel furnace was the first to be developed and is illustrated in *Figure 7.8*. It consists of a large upper bath with a small channel at the bottom of the furnace. Heat is induced in the channel via a coil and an iron core surrounded by a ceramic refractory, which permanently contains molten metal. It is the most energy efficient of the induction furnaces, but is only economic if run continuously, with molten metal always maintained in the bottom channel. If production reverts to single shift, the power still has to be left on. Thus it is not very suitable for the trade die caster who handles a range of alloys and who does not operate a three shift system.

Molten aluminium contains minute particles of aluminium oxide which has about the same density as the pure metal; these particles tend to move to the side of the inductor channel, causing clogging which introduces a serious technical difficulty in the channel furnace. Over the past decade, attempts have been made to overcome this handicap, including a method of injecting nitrogen, but the problem has not been solved completely.

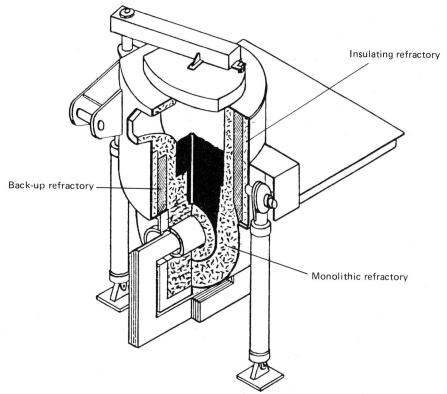

Insulating refractory

Back-up refractory

Monolithic refractory

Figure 7.8 Core or channel-type furnace with internal melting loop surrounding the iron core

Furthermore, maintenance of channel furnaces is difficult and, if the channel has to be emptied, it must be done by removing bricks and clearing the metal from underneath or by tilting the furnace to an abnormally high degree. Therefore the recent tendency in casting has been to prefer coreless induction furnaces, but channel furnaces are suitable for large scale melting, for example in the manufacture of aluminium alloy ingots, where furnaces have a capacity of 25 tonnes or more. Among other uses, they feature in aluminium electrolytic smelting plants[6].

Coreless induction furnaces

In this type of induction furnace, now becoming widely used for aluminium alloy melting, electric current is induced in the metal, leading to the evolution of heat[6] and hence to melting. A water cooled primary coil of copper tube, to which alternating current is applied, surrounds a melting chamber or crucible as illustrated in *Figure 7.9*. The induced current causes the metal to melt and sets up a stirring action around two concentric centres in the upper and lower parts of the molten bath. It is possible to melt ingot, scrap, turnings and normal foundry returns with lower metal losses than are obtained in other furnaces.

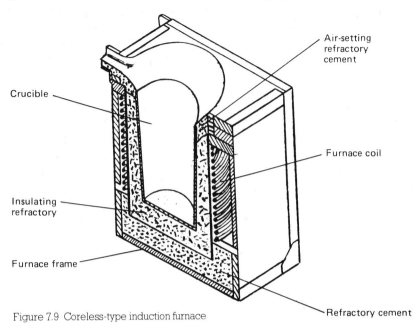

Crucible

Air-setting
refractory
cement

Furnace coil

Insulating
refractory

Furnace frame

Refractory cement

Figure 7.9 Coreless-type induction furnace

The primary current-carrying coil, made of thick-walled copper tube, is in the form of a helix, placed in position in the furnace casing and securely held in position with tie bars. The coil is first coated with a refractory slurry, which hardens and assists in keeping the coil rigid. A refractory cement is rammed into the gap between the inner wall of the coil and outer wall of a removable former; the crucible shape made in this way is fired to the required temperature, usually by applying the furnace power to a susceptor placed inside the crucible before introducing the first material. Alternatively, the melting chamber can be provided by a preformed ceramic, or silicon carbide crucible, depending on size of furnace and operating frequency.

During the past 20 years, the power-conversion efficiency of induction furnaces has doubled. Until about 1960 spark gap systems were used, with a power-conversion efficiency of about 50%. Next, motor generator sets were developed, increasing the efficiency to about 75%. In the late 1960s solid state electronic systems increased the power-conversion efficiency to about 94%, leading to a metal melting efficiency of about 45%. During the early 1970s capacitor switching of the many silicon controlled rectifiers was needed, but by 1974 the developing technology of small integrated circuits made it possible to eliminate capacitor switching during a melt.

A wide range of frequencies is used for induction melting. For large furnaces of over 5 tonne capacity ordinary 50 Hz power is suitable. For lesser tonnages, the range of available frequencies is 250, 500, 1000, 3000 and 10 000; the smaller the amount of metal to be melted, the higher the frequency. For a typical die casting plant a 1000 Hz, 400 kW furnace, melting about 700 kg/hour, is appropriate. In the 1000 Hz range, a 150 kW furnace melts about 225 kg aluminium per hour, while the largest furnace in the range, with a power requirement of 1500 kW melts a maximum of 2700 kg/hour.

These furnaces are lip axis tilted for pouring. The furnace shown in *Figure 7.10,* was installed in a British die casting plant, where it replaced two reverberatory furnaces,

Figure 7.10 Coreless induction melting furnace. (Courtesy Inductotherm Ltd., and Brockhouse-Kaye Ltd.)

providing lower melting cost, lower metal loss and a higher standard of metal quality. The control cabinet shown in the illustration was installed for the first purchase but its controls and water cooling facility are appropriate for two more. Thus the installation of subsequent furnaces was not attended by the need to purchase a control cabinet and equipment, which made up about one third of the initial cost. In a normal cycle, the coreless furnace is emptied except for a heel remaining at the bottom of the crucible. It is then charged with ingot and scrap; the current is switched on and its power adjusted so that melting rate is as high as possible. When the metal is melted the power is reduced and then switched off when pouring is about to start. Following weekend breaks in production, the coreless induction furnace can be restarted from cold by an automatic system which regulates the power supply.

Channel-less detachable inductor furnaces

In this equipment, an inductor containing the power coil and refractory can be separated from the upper chamber containing the bulk of the metal. It is a comparatively recent development in the melting and holding of aluminium alloys, used at present in small bale-out furnaces. It works in the same way as a channel furnace, but with all the advantages and flexibility of the coreless furnace. The melt can be poured from a lip axis spout if the melt chamber is to be emptied completely and, unlike the conventional channel furnaces, at initial start up, the furnace can accept a starting plug of metal inside the coil and melt from cold, whereas the channel furnace always has to be primed with molten metal. Recent trials show that the economics of the channel furnace can be achieved without the disadvantages of a fixed alloy melting programme. It is expected that the future will see wider use of the detachable inductor furnace in the die casting foundry, where its ease of maintenance will surely be welcomed.

Transport of molten metal to the die casting machines

Having provided a central source of molten metal, appropriate supplies must be conveyed to each machine. Many furnaces including basin tilters and some reverberatory furnaces, can be tilted for metal removal; some furnaces have tap-out valve assemblies for metal pouring. Pumping aluminium with immersed pumps, although satisfactory for zinc and lead alloys, is not suitable for aluminium, owing partly to the higher temperature but mainly to the tendency of molten aluminium to dissolve ferrous components.

A simple but effective method of metal transfer which has been in use for more than a decade in some die casting plants relies on air pressure, in a similar manner to the air pressure metal transfer systems described on page 169. Molten aluminium alloy is poured from a tilting reverberatory furnace into a travelling ladle on a fork lift truck. When the metal-carrying truck arrives at the first machine, the cover, which has been in position above the ladle, is lowered and clamped securely, making a pressure tight seal. The cover is connected to an air pump located on the 'roof' above the truck driver's head. He switches on the pump and pressure is applied to the surface of the aluminium which flows through a refractory-lined tube into the holding furnace of the cold chamber machine.

There are several other ways of transferring the molten aluminium from the bulk melter to the holding furnace. Crucibles travelling on monorails can be used where the production is controlled to a fairly stable pattern, as for example a motor manufacturer making large quantities of a limited number of die cast components.

Where versatility is required due to ever-changing production, or where a variety of alloys must be handled in any one area, driver-controlled fork lift trucks can be used for carrying transfer ladles. The molten aluminium can either be pumped from the ladle to the holding furnace as described above or the fork lift mechanism is used to lift and tilt the ladle to dispense the molten metal to each machine. By utilizing the truck's own power system, the ladle unit can be manoeuvred precisely; full tilting facilities provide a controlled transfer of molten metal and, after a short training, drivers can regulate the amount of metal poured with safety and accuracy. Quick disconnection facilities will allow the truck to be used for handling solid materials between intervals of molten metal transportation. When required for travelling ladle operation, the crucible and ancillary hydraulic or electrical systems are connected and locked in position. For manufacturers who do not wish to use fork lift trucks, specially designed mobile ladles are available. They can carry about 200 kg of molten aluminium and can either be manually propelled and tilted, or battery driven.

One of the problems of organizing a completely mechanized aluminium die casting plant arises from the need to provide supplies of molten metal at the right time, at the right place and in the right amounts for each machine. This is always difficult for a trade die caster whose daily programme is dependent on the schedule changes and special urgencies of many customers. Even the most highly mechanized plants rely on simple methods of signalling when crucibles are ready to take further supplies of molten metal. The driver of the fork lift truck transporting molten metal, patrols the foundry and from his knowledge of the jobs in production he is aware when furnaces need replenishing.

Some foundries have signalling lights to alert the truck driver. When full automation eventually comes to the aluminium die casting plant, it will be necessary for an automatic signal to be given when metal drops below a certain level.

Melting-holding furnaces

Like many other processes in Nature and in industry the alteration of one feature of the die casting process sets up a train of new conditions, some of which may not have been anticipated. Before the automatic production of aluminium alloy die castings, a manual operator could move 100–150 kg of molten metal per hour and a medium sized bale-out furnace adjacent to the machine was capable of melting the required amount of metal. When the machine was mechanized, the rate of output increased two or three–fold and the fossil fuel fired crucible furnace was no longer able to provide sufficient molten metal. This emphasized the need either for a central bulk melter feeding regular supplies to the bale-out holding furnace, or a melter-holder adjacent to the machine large enough to provide ample molten metal. It also emphasized the need to make furnaces more efficient.

Immersed crucible furnaces, developed during the 1970s, have been described in the section on bulk melting, but many die casting plants are using them for providing a combination of bulk melting and bale-out. Such furnaces do not normally have a tap-out system because they are intended for removal of the metal by automatic ladling. Small reverberatory furnaces used as melter-holders often incorporate a system of double chambers where the charge is melted in one section of the furnace and clean, molten metal is ladled from an adjacent side well. Bale-out crucible furnaces are normally used for holding, and as melting furnaces at the die casting machine stations; recently their efficiency has been improved with heat recuperation systems, as discussed below. The use of channel induction furnaces has declined due to the technical problems discussed earlier, but the new channel-less detachable inductor furnaces are expected to find some application as melter-holders.

Furnace efficiency

Although die casters and others in the foundry industry of the early 1960s probably suspected that their furnaces were only about 10–15% efficient, they had not realized how much could be done to make substantial improvements in thermal efficiency. Indeed, managements were more concerned with production problems than with possible savings of energy, which then appeared to be of limitless supply. When the cost of energy began to increase, foundrymen and furnace manufacturers investigated how heat was being lost and what could be done to prevent the loss. Over 60% of the furnace heat escaped with the exhaust gases; 15% was lost through the furnace lining and less than 25% of the available heat was used for melting the metal. The efforts to make furnaces more efficient have been directed in two ways. Firstly the vast amount of exhaust heat can be used to preheat the air drawn into the burner, thus saving fuel. The wastage of heat through the furnace lining has been tackled by the use of ceramic fibres, discussed later.

Hot flue gases can be passed through a recuperator which preheats incoming combustion air to temperatures up to about 350°C. This system can be used on all fossil fuel

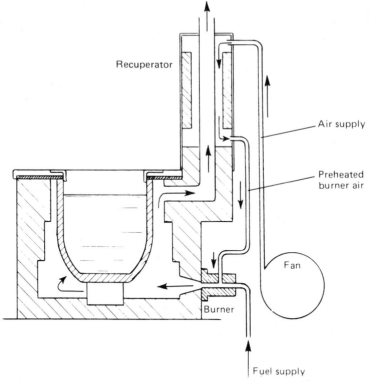

Figure 7.11 Diagram of high efficiency bale-out furnace. (Courtesy Morganite Thermal Designs Ltd.)

fired furnaces but maximum economies are achieved when applied to bulk melting or melting-holding furnaces under conditions of high metal throughput and continuous burner operation. *Figure 7.11* illustrates the general principle of the system on a crucible furnace. The incoming air passes down the outer section of the flue and is preheated by the exhaust gases[7]. Potential fuel savings depend on the temperature and volume of flue gases wasted, while the actual saving depends on how much of this heat is usefully recovered to preheat incoming burner combustion air.

Holding furnaces

The bale-out crucible furnace, either gas, oil fired or electric is still widely used. The crucible materials are either silicon carbide or clay-bonded graphite, since aluminium attacks and dissolves iron. Although other materials have been tested, none of them have been found to provide the combination of comparatively low cost and long life offered by these two materials. Clay-bonded graphite crucibles cost less than those of silicon carbide but their thermal efficiency is considerably lower and their performance deteriorates over a period. Nowadays only about a third of all holding furnaces and only about a tenth of all bulk melting units have clay-graphite crucibles.

Before foundries in general, and die casters in particular, became metallurgically and energy conscious, a typical crucible furnace would consist of an outer steel shell lined with refractory into which a crucible was placed. The hot combustion gases circulated round the crucible and melted the metal by heat conduction. Such equipment required 2½ million kcal/tonne. Since theoretically 1 tonne of aluminium requires only about 250 000 Kcal/tonne, the efficiency of the older furnaces was only about 10%. A breakthrough came in about 1960, when the type of packaged burner used to fire central heating installations was applied to bale-out furnaces. Early designs required a supply of compressed air to aid atomization but eventually this was eliminated by improved burner design. Firing into the restricted combustion space in a bale—out furnace involved different problems from those of boiler firing. A special flame shape is needed and back radiation

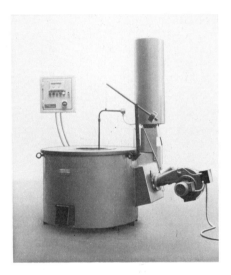

Figure 7.12 Oil fired crucible bale-out furnace. (Courtesy Morganite Thermal Designs Ltd.)

on the burner is much greater. *Figure 7.12* shows a modern bale-out furnace with self contained burner unit and automatic metal temperature control.

Electric resistance holding furnaces

Changes in the relative prices of electricity versus gas and oil have brought the cost of energy for operating electric resistance furnaces close to that of fuel fired ones, and they have the added benefit of better working conditions. Earlier designs of electric resistance furnaces suffered from the disadvantage that elements were liable to premature failure, resulting in the loss of one third or more of the power input, and long delays occurred while failed elements were being replaced. Improved designs have been developed, a typical arrangement being shown in *Figure 7.13*. Here the elements are arranged in a hexagon around the crucible and are supported by refractory pillars. To replace a heater assembly of elements and refractory panel, only the top cover of the furnace needs to be removed; the element is then disconnected at the terminal blocks and the assembly withdrawn from the support pillars. As the heater elements are connected in parallel, failure of one element results in only a small reduction of heat input; the replacement of a failed element can be left until a suitable break in production[8].

In the past, electric resistance furnaces were used mainly for holding; they were fed with molten metal and the operator baled out the metal for each shot. New designs are

Figure 7.13 Morgan electric resistance crucible bale-out furnace

now available for melting at rates approaching that of the fossil fuel fired furnaces. Resistance furnaces have several advantages: they require less maintenance, there is no burner quarl to keep in correct shape and no oil filter or burner to clean.

Electric radiant holding furnaces

Increasing use is being made of electric radiant furnaces for holding aluminium as a result of substantial energy savings which have been achieved. The secret behind their high efficiency is the extensive use made of thermal insulating materials. *Figure 7.14* illustrates a furnace of Swedish design which uses a thick layer of insulation within its outer steel case. A shallow metal bath in the centre of the furnace contains overhead heating elements giving direct radiation onto the melt surface. Adjacent charging and take-out chambers are separated from the heating bath with walls incorporating small ports at their base to allow metal to flow within the sections. Special nonwetting hot face insulating materials contain the melt. A 380 kg capacity holding furnace for aluminium has a maximum connected power load of 9 kWh with an operating consumption as low as 3 kWh during production and 2 kWh during non-production hours, when the take out section is covered with an insulated lid.

Furnace insulation

Ceramic fibres are used either in conjunction with refractory bricks or as a complete replacement. Their very high insulating properties contribute to reductions in furnace energy costs. Since they have only about one-third the heat conductivity of firebrick they may be installed in correspondingly thinner layers for a given heat loss. Based on alumino–silicates, consisting of about 45% alumina, 52% silica with oxides of iron and titanium, these materials have an ability to withstand thermal shock, and they do not suffer failures by spalling and cracking during rapid heating and cooling. A high percentage of the volume is made up of small, discontinuous air-occupied voids which reduce heat transfer. Infra-red radiation transmission in those insulating materials is reduced by the metal oxide additions.

These ceramic materials are normally supplied in the form of a blanket, compressed blocks or as fibres; they can also be vacuum formed into definite shapes[7] or mixed with small additions of cement and other chemicals and cast into the required shape. In holding

Figure 7.14 Electric radiant holding furnace. (Courtesy Ugns and Gasteknik)

furnaces blocks of the material may be primed or glued directly to the inside of the furnace casing. Alternatively the hot face lining is cast in place and the insulating material, in fibre or blanket form, is used as an in-fill material between the cast lining and the steel furnace case. There are some disadvantages. The ceramic fibres are not very strong, and they cannot be installed on furnace floors or in other areas where they would be subject to abrasion. Additionally, since their insulating properties depend on a high porosity, they should be used with caution in situations where the pores might become plugged by deposits from fumes.

References

1.   (Editorial)
      The molten aluminium story. *Metal Bulletin,* No. 4751, p.iii (November 30th 1962)

2.   THORNTON, J.
      Molten aluminium shipping stirs differing views on energy use. *American Metal Market,* **85,** 70
      (May 2nd 1977)

3.   KENNEDY, S.F.
      Fossil fuels or electric – how should you melt? *Society of Die Casting Engineers* Congress Paper
      No. G– T79–075 (March 1979)

4.   Instruction pamphlet: *Morgan Immersed Crucible Furnace,* Morgan Thermal Designs Ltd
      No. FDI 703

5.   ROWAN, H.M.; HORVATH, G.
      The role of multiple frequency induction melting systems. *45th International Foundry Congress*
      Paper No. 31 (Budapest 1979) Inductotherm Eurpoe Ltd., Worcestershire

6.   SCHAUB, H.P.
      Induction furnaces for melting and holding aluminium. *Foundry Trade Journal* **134,** (2936),
      331 (March 15th 1973)

7.   ATKINS, R.
      Developments in metal melting and holding in non-ferrous foundries. *The British Foundryman,*
      74, 161–172

8.   ATKINS, R.
      Recent developments in electric melting and holding for non-ferrous foundries. *Foundry Trade
      Journal,* **151,** 270–282

Chapter 8

Inclusions and hard spots in aluminium alloys

The term 'inclusion' refers to every type of undissolved foreign material present in a cast metal; 'hard spots' specifically identify forms of inclusion which have a harmful effect during machining, causing rapid wear, and sometimes breakage, of cutting tools. A machined aluminium surface containing hard spots shows small bright or dark areas; often the included particles are dragged from the surface, leaving irregularly shaped voids.

Some inclusions are inherent in die casting alloy raw materials. Oxides originate from the alumina in the electrolytic cells in which primary aluminium is produced. Research conducted in Norway[1], identified numerous inclusions in commercial grades of the primary metal. These contained 6–16 ppm of oxide, 3–12 ppm of aluminium nitride, 1–2 ppm of Al_4C_3 and less than 1 ppm of cryolite and borides. However most of the inclusions in die castings are formed during the melting and holding of molten metal before it is cast. The development of mechanized and automated die casting processes and high speed machining of components has made it increasingly necessary to avoid inclusion problems by good foundry practice and housekeeping. There are several types of inclusions encountered in the die casting plant. They are listed below, showing the most frequent and deleterious first.

1. Aluminium oxide, either as a skin of oxide formed on the molten metal during transport, or particles of aluminium oxide produced during melting and holding.
2. Intermetallic compounds formed firstly by the metal composition creating the right conditions and secondly by allowing them to segregate through inefficient temperature control or by neglecting to stir the metal in the crucible.
3. Non-metallic particles accumulated in the metal from ladles, crucibles or tools.
4. Small particles which have formed during the injection of the die-cast metal and have solidified separately.

Aluminium oxide

Although solid aluminium has a thin film of oxide which is protective, the molten metal forms oxide at a rate which increases with temperature; this is a major cause of hard spots. Any disruption of this film on the surface of the metal from turbulence during pouring or excessive disturbance when ladling will lead to a continuous build up of further oxide formation. The density of aluminium oxide is not significantly different from that of the molten alloy and consequently the oxide particles remain suspended in the melt and are transferred to the casting. Sometimes oxide skin sticking to the pouring ladle after the casting operation is inadvertantly returned to the melt. Oxides may act as nucleation centres so that primary crystals of high melting point compounds may be observed in this kind of inclusion. *Figure 8.1* shows a group of inclusions in an aluminium—

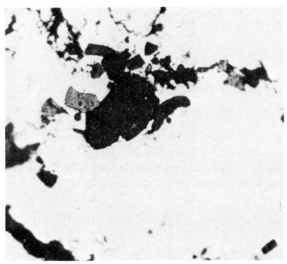

Figure 8.1 Aluminium oxide and silicate inclusions in an aluminium – silicon – copper alloy pressure die casting × 500 (Courtesy Alcan International Ltd.)

silicon—copper alloy die casting. The main constituent is α Al_2O_3 (dark grey area) associated with a smaller amount of aluminium silicate (pale grey area).

It has been reported by M. Leoni and F. Fommei[2] that, in the presence of molten aluminium, the oxide skins on the melt surface which originated in metal transfer and ladling undergo a transformation from their transition structure of η alumina Al_2O_3. This phenomenon is believed to have originated from a different solubility of the η and α phases in molten aluminium, which causes η Al_2O_3 to dissolve and subsequently reprecipitate as α Al_2O_3, a harmful type of inclusion with high hardness and comparatively large size. In both cases, the solubility values involved are extremely low. The phenomenon can be observed only after the melt has been held for a period of time, and the resultant castings examined. Holding furnaces used on a continuous operation can encourage the incrustation of α Al_2O_3 with intermetallic compounds which slowly sink in the melt.

The harmful effect of aluminium oxide was realized more than 20 years ago and, among several useful papers, one by F.H. Smith[3] gave guidelines for minimizing the formation of oxide inclusions. He pointed out that the higher the proportion of foundry scrap used, and the smaller the pieces, the greater is the amount of oxide introduced into the melt from the surface film on the solid metal in the charge. The presence of moisture and oil promotes further oxidation, so only clean foundry scrap should be used.

Intermetallic compounds

Iron, manganese and chromium in aluminium–silicon alloys form high melting point, hard, intermetallic compounds which cause excessive machine tool wear. When these elements are present in sufficient amounts and when melting and metal treatment conditions are not controlled efficiently, the intermetallic compounds are formed around solidification centres provided by aluminium oxide nuclei present in the molten metal. Being heavier than the aluminium, the solidified intermetallics sink to the bottom of the melt, causing a sludge to form. The rate of precipitation is temperature dependent, although if any of the compounds have already formed in the ingot material as supplied, sludge formation occurs more rapidly. Maintaining too low a temperature (generally in holding furnaces) encourages their precipitation and the longer the metal is retained in the crucible the more the segregation. Under these conditions there is the hazard that a ladle immersed deep into the melt will collect these inclusions which consequently become entrapped in the die casting, leading to difficult machining and poor surface finish.

As long ago as 1950 the composition of the sludges formed from these intermetallic compounds was being investigated. D.L. Colwell and O. Tichy gave a paper to the American Foundryman's Society[4] which was then given wider circulation through Modern Casting[5]. An aluminium alloy, A.380, similar to LM24, was held for several hours at about 680°C. *Table 8.1* shows the percentage composition of the metal at the top and bottom of the crucible.

TABLE 8.1

	Si	Cu	Mg	Fe	Zn	Mn	Ni
Top	9.35	3.45	0.20	0.95	0.96	0.35	0.19
Bottom	10.10	2.96	0.07	5.0	0.01	2.0	0.64

When the metal was thoroughly mixed after melting and held at 700°C for a short time the percentage composition of the top and bottom of the bath were similar, as shown in *Table 8.2*.

TABLE 8.2

	Si	Cu	Mg	Fe	Zn	Mn	Ni
Top	8.75	3.38	0.04	0.90	0.87	0.32	0.22
Bottom	8.50	3.33	0.04	0.87	0.83	0.33	0.22

In a paper by R.P. Dunn[6] information was reported about the chemical composition of sludges in the alloy A.380. The material removed from the bottom of a melt gave an analysis which included 6.5% Fe, 3.03% Mn and 1.44% Cr. The actual crystals of the compound in the sludge were analysed by microprobe, giving a composition which included 20% Fe, 10% Si, 6% Mn, 1% Cr and 1% Cu. Another analysis was made of large crystals from sludge which had accumulated over a long period in the charging well of a reverberatory furnace melting A.380 alloy. This sludge had a melting point of over 800°C and contained 9.4% Si, 11.38% Fe, 6.07% Mn, 2.16% Cu, 1.77% Cr and 0.4% Zn.

Figure 8.2 and *Figure 8.3* illustrate the intermetallics that could be formed by prolonged 'stewing' of the aluminium–silicon–copper alloy. The melt was held at 660°C for 8 hours, allowed to solidify and then examined. *Figure 8.2* shows the appearance of

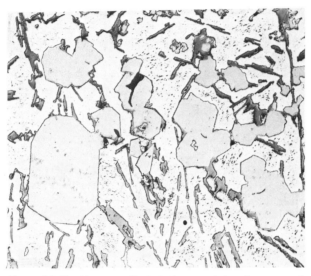

Figure 8.2 Hard intermetallic inclusions consisting of segregated Al(FeMn)Si constituents in an aluminium–silicon–copper die casting alloy 5 cm from the crucible surface. × 200 (Courtesy Alcan International Ltd.)

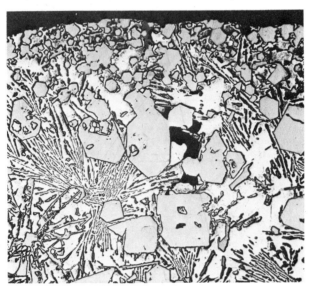

Figure 8.3 Hard intermetallic inclusions consisting of segregated Al (FeMn) Si constituents in an aluminium–silicon–copper die casting alloy adjacent to the crucible surface. ×100 (Courtesy Alcan International Ltd.)

metal taken from about 5 cm above the bottom of the crucible. The large angular light grey crystals are intermetallic compounds of Al(FeMn)Si. The needle formations are silicon and the spots $CuAl_2$. *Figure 8.3* is a photomicrograph of the metal taken from the bottom of the melt adjacent to the crucible surface where segregation was more pronounced and led to heavier sludging. It is not likely that such a vast accumulation of intermetallics would be found in an aluminium alloy in a die casting plant but these illustrations show the ultimate in metal maltreatment.

Several empirical formulae have been developed to predict whether a given alloy will be more likely than usual to cause intermetallic compounds to be formed and to segregate. One formula recommended by the Society of Die Casting Engineers[7] is.

%Fe plus (1.5 x %Mn) plus (2 x %Cr) = 1.85 max

For example, a melt of A.380 containing 0.9% Fe, 0.6% Mn and 0.1% Cr would have a total 'sludging tendency' of 2.0 and such a composition would be more liable to the formation of sludge than an alloy with 0.9% Fe, 0.3% Mn and 0.05% Cr, which would have a 'sludging tendency' of 1.45. The formula indicates the cumulative effect of iron, manganese and chromium in the ratio of their influence (1:1.5:2).

In the paper by R.P. Dunn[6], mentioned above, a different formula was proposed, linked with the temperature of the metal. The sludging factor was calculated by adding Fe % plus (2 x %Mn) plus (3 x %Cr). In an alloy with 1.0% Fe, 0.3% Mn and 0.1% Cr, the sludging factor would amount to 1.9 and this would correspond to a permitted holding temperature of 650°C. In an alloy containing slightly lower impurities, 0.8% Fe, 0.3% Mn and 0.1% Cr, the sludging factor would work out at 1.7 and the minimum temperature could be lowered to 625°C.

Some years before the American investigations, useful research was done by the then British Non-Ferrous Metals Research Association and reported in Metallurgia[8]. The die casting alloy, then known as LAC 112, similar in composition to the British LM2, and American A.384.1 was studied with relation to iron and manganese contents, linked with the type of die casting and the operating conditions, as shown in *Table 8.3*.

TABLE 8 3

Use	Iron–manganese formula
Heavy section castings, melts being kept hot enough to avoid segregation before pouring	Fe% plus 1.5 Mn% = 1.5
Melt to be held at 650°C after heating to well above this temperature	Fe% plus 1.5 Mn% = 1.4
Limits for avoidance of segregation under most severe conditions	low silicon: Fe% plus 1.5 Mn% = 0.9 high silicon: Fe% plus 1.5 Mn% = 0.75

It is advisable that die casting manufacturers discuss their requirements and acceptable impurity limits with their ingot suppliers. It must however be emphasized that the 'sludging formulae' indicate the levels above which there is an enhanced tendency to form intermetallics, but careless metal control can still cause these inclusions to accumulate to a lesser extent. Once the sludge is formed it will not easily redissolve in the aluminium, even if it is left for long periods. Careful stirring is necessary in conjunction with an increase in temperature but it is far better to take the action that will avoid the production and segregation of intermetallic compounds.

Prevention of oxide and intermetallic segregation

The sources and causes of inclusion contamination arise during all the stages from the manufacture of the ingot to the final injection of that metal into the die. It is desirable that the metal is obtained from a supplier with the resources to obtain ample good quality secondary and primary metal, to refine that metal efficiently and to be cooperative in agreeing specification limits that will provide the best opportunity of making die castings free from inclusions. If the metal is delivered molten, the possible formation of oxide should be understood and the treatment of the metal from the time of its arrival to its distribution into the foundry should be controlled accordingly. It is not difficult to obtain the analysis of the elements in a supply of metal, but it is more difficult to measure the amount of aluminium oxide entrapped in that metal. Oxide segregations do not allow for representative samples to be obtained, while wet chemical analysis methods are time consuming.

The choice of bulk melting equipment is discussed on page 53, but die casting managements have to make a decision between the undoubted speed of melting in reverberatory furnaces against the reduced risk of metal deterioration in the slower operating crucible or electric furnaces. Melting furnaces should provide the highest throughput capacity with the smallest possible area of contact between liquid metal and furnace atmosphere. These features will also reduce oxidation, which is both temperature and time dependent. All melts should be well stirred throughout their depth as soon as all of the metal is molten, preferably having a surface protected by a covering flux which will also absorb many of the impurities present. Degassing agents can also assist in cleaning molten metal, to bring suspended non-metallic inclusions to the surface where they can be absorbed into the flux cover. In addition, agitation of the melt towards the bottom of the furnace discourages intermetallic segregation. All melts treated in this way should be allowed time for suspended particles to settle out before dispensing any of the melt to the die casting station.

Following diligent control during melting, further precautions must be taken while transferring the metal, first to the holding furnace and subsequently into the die, to avoid excessive turbulence and to ensure that no foreign matter accompanies the metal. It is necessary that the molten metal should be taken from below the melt surface. Ladles and ladling technique must be selected to prevent turbulence. A die caster manually ladling the metal could prevent inclusions from being withdrawn but some early mechanized ladling systems did not reproduce the skilful action of the manual operator, so inclusions were entrapped.

Crucibles should be maintained as full of metal as possible and the level of metal should never be allowed to fall below 30 cm from the bottom. Nowadays, most melting and holding equipment is provided with automatic temperature control but it is necessary to establish a program of regular pyrometer calibration. Furnace areas should be maintained clean; crucible walls need scraping clean, preferably on a regular daily basis, but at least weekly. Turbulence must be minimized at all stages of alloy melting, holding and pouring to reduce oxide formation.

Oily and contaminated scrap returns need to be withheld for separate melting. Very thin castings, flash and thin sections produced in the die casting process, should be treated separately. Even in the best regulated plants, breakdowns and delays occur; if the hold-up is only temporary, foundry management and tool room personnel are involved in doing minor repairs or checking the operating conditions. Under such circumstances it is easy to forget that the molten metal has been remaining in the crucible, collecting aluminium oxide if the temperature is too high or forming sludge if the temperature has dropped. Therefore after a halt in production the metal may have to be cleaned and stirred.

In the past, some die casting foundries maintained metal levels in holding furnaces with the continuous addition of ingots, scrap or slugs from the casting shot. This introduced the conditions under which intermetallic segregation occurred, leading to customer complaints − particularly during the late 1960s, when high speed machining plants were being installed by large users of die castings. Consequently the problems caused by hard spots became much more serious than in earlier years when machining processes were not automated. Fortunately the realization that something had to be done about these machining problems was followed by improvements in metal melting furnaces, better metallurgical control in the die casting plants and an almost complete stop in the practice of returning solid metal to the holding crucibles. When central melting facilities are not available, as for example in small non-mechanized plants, the prevention of inclusions can be reduced by employing two furnaces so that metal is being melted in one furnace and then transferred to an adjacent furnace.

Problems of automation

With the introduction of automatic ladling systems, attention must be given to ensure that oxide pick-up from the surface of the melt is prevented, and the liquid surface is not disturbed unduly. When mechanized ladles were first developed, insufficient attention was given to the correct flow of the molten metal from the ladle to the die, and casting quality suffered accordingly. The pouring mechanism should be designed so that the distance from the lip of the ladle at the pouring position and the opening in the die casting machine's injection cylinder are as short as practicable. Several suitable automatic pouring mechanisms are discussed and illustrated in Chapter 18.

Mechanization is doing so much to eliminate manual labour that the need for good metallurgical housekeeping is sometimes overlooked. Automatic ladling systems have the disadvantage that once the mechanism is set up, skimming of the metal surface is not so easy as when the crucible was more accessible. On the other hand, as oxidation of the metal increases with time, the rapid production of the automatic machine with its mechanized pouring system gives less opportunity for oxidation to occur.

Other causes of inclusions

Sometimes non-metallic inclusions are introduced accidentally into the molten metal as refractory particles from furnaces or crucibles or decomposed products from lubricants and excessive use of fluxing materials. *Figure 8.4* shows dark grey clusters of an aluminium–magnesium spinel oxide $MgAl_2O_4$ in an aluminium–magnesium alloy.

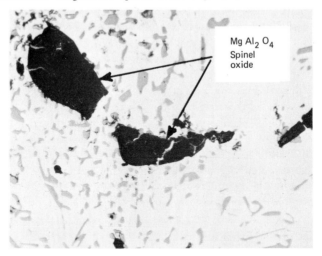

Figure 8.4 Clusters of an aluminium magnesium spinal oxide in an aluminium–magnesium die casting alloy. × 500 (Courtesy Alcan International Ltd.)

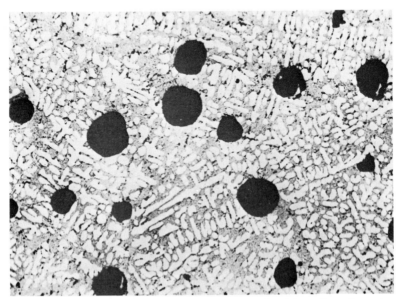

Figure 8.5 Porosity voids caused by hydrogen gas in an aluminium–silicon die casting alloy. × 200 (Courtesy Alcan International Ltd.)

Formed by reaction between silicates or refractory materials and the melt, they are often contaminated with impurities such as silicon, iron, potassium, sodium and/or calcium. When metal levels in the holding furnace have been allowed to fall too low, overheating of the upper parts of the crucible wall occurs and both the adhering metal oxides and residual fluxing products will burn and attack the crucible and refractory coating to form hard inclusions which dislodge and enter the molten alloy. Such defects can be prevented by a sensible programme of maintenance in which ladles and crucibles are examined regularly and cleaned or replaced when necessary.

The die casting management should also be alert for unexpected hazards introduced by new methods. For example, if processes are being developed for barrel degating and deflashing, linked with a conveyor system, sooner or later some of the media used in the barrelling may find their way to the metal melting section and cause unexpected hard spots.

The fourth cause of inclusions listed on page 71 results from die design and gating problems. If the injected metal does not freeze as one entity and small spots of metal become separated from the bulk of metal in the first stages of injection, those particles may remain embodied in the casting, causing tool wear in machining processes.

There are other causes of machining difficulties, though they cannot be classified as inclusions. Voids in a porous die casting cause tool chatter, leading to premature break-down. *Figure 8.5* shows an aluminium−silicon alloy with voids in such profusion that machining problems would undoubtedly occur.

References

1.   SIMENSEN, C.J.; BERG. G.
      A survey of inclusions in aluminium. *Aluminium,* 56, (5)

2.   LEONI, M.; FOMMEI, F.
      Causes of hard spots in pressure diecastings, and technological remedies. *Alumino,* 5, 246−252
      (1976) (in Italian and English)

3.   SMITH, F.H.
      Inclusions in Aluminium Alloy Castings. *Light Metals,* 123, 208−212 (August 1960)

4.   COLWELL, D.M.; TICHY, O.
      Machinability of aluminium die castings. *American Foundryman's Society,* Preprint No. 56−65
      (May 3−9, 1956)

5.   COLWELL, D.M.; TICHY, O.
      Once there was a hard spot in an aluminum die casting. *Modern Casting,* 30, 28 (July 1956)

6.   DUNN, R.
      Aluminum melting problems and their influence on furnace selection. *Die Casting Engineer*
      9, (5), 8−30 (September−October 1965)

7.   BRUNER, R.W.
      Metallurgy of Die Casting Alloys. An engineering technology textbook. Society of Die Casting
      Engineers, Detroit (1970)

8.   GLAISHER, W.H.
      Segregation of iron and manganese in some aluminium casting alloys. *Communication from
      BNF Metals Technology Centre* (Then British Non−Ferrous Metals Research Association),
      Report R.R.A. 905P, (March 1951). Reported in *Metallengia* 43, (257), 127 (March 1951)

Chapter 9

Zinc and zinc alloys

In the late 1970s, the annual free world production of zinc, amounting to about 4½ million tonnes, was exceeded by only three other metals: iron, aluminium and copper. The UK consumed about 250 000 tonnes. The major uses of zinc are divided into about 40% for anticorrosion coatings for steel, 20% for die casting, 20% for making brass, 10% for rolled zinc and the remainder for zinc dust, pigments and miscellaneous applications.

There is a wide distribution of zinc ores throughout the world, though all of the deposits are comparatively lean, averaging about 4% metal content. The major mines, in order of annual tonnage, are in Canada, the USSR, Australia, Peru and USA. Recently zinc ore deposits have been found in Eire, and mines have been developed in Greenland[1]. Zinc can be produced economically from a number of ores but nowadays zinc blende, containing the sulphide, often associated with lead and cadmium compounds and sometimes with silver and mercury, provides most of the world's zinc output, the remainder coming from calamine, a zinc carbonate. The deposits are mostly worked by underground mining and are concentrated by flotation near the mines. A considerable amount of the concentrate is taken to Europe, Japan and USA for extraction of the metal.

When zinc sulphides arrive at the smelting plant, they are roasted to form zinc oxide and sulphur dioxide, which is then converted to sulphuric acid, some of which is sold and some used in the electrolytic refining of zinc. The metal is obtained either by thermal processes developed from the early methods of zinc smelting or, more commonly, by electrolytic processes. Nowadays about four-fifths of the total zinc production is obtained by electrolysis[2]. Zinc oxide is dissolved in dilute sulphuric acid and the zinc sulphate solution so formed is treated chemically to remove impurities. Next the purified solution is pumped into electrolytic cells which have aluminium cathodes and lead/silver anodes. At intervals, ranging from 24 to 72 hours, zinc is stripped from the cathodes, melted and cast into ingots. The deposited metal has a purity of at least 99.95% and can be maintained above 99.99% if required. The success of zinc die casting alloys depends on the availability of 'four nines plus' zinc; they will be unreliable if they contain more than a few g/tonne of lead, cadmium or tin.

The consumption of electrical power per tonne is about one-quarter that required for the electrolytic smelting of aluminium. Taking into account the fact that the density of zinc is about 2.4 times greater than that of aluminium, the energy required for smelting a unit volume of zinc is 40% less than that for aluminium.

Thermal process

In the early years of zinc production, thermal processes were used, based on William Champion's development[3] of a vertical retort in 1738. A temperature of about 1000°C is required before zinc oxide can be reduced by carbon to the metal. However zinc vaporizes at 906°C and therefore must be condensed back to liquid metal before it comes into contact with air, which would oxidize it. Champion's method was to roast a zinc carbonate ore, making zinc oxide. This was put in a lidded crucible in layers with coke. A crucible with a hole in the base was placed over an iron tube leading to a cool chamber below, where another crucible contained water into which the zinc vapour condensed.

Technical improvements in Belgium and England during the 19th century led to a method of smelting in horizontal retorts and the zinc was collected in clay tubes fitted to the end of the retorts. Until the early 1950s half of the world's zinc was produced by methods based on these early processes. Then a radically new development led to the announcement in 1957 of the Imperial Smelting Process[3] which was more flexible than the earlier processes, allowing the treatment of low grade and high grade zinc ores as well as mixed zinc–lead ores.

Roasted zinc and lead concentrates are charged into a blast furnace with a pro-portioned amount of coke; a preheated blast of air is injected through water cooled tuyeres. Slag and molten lead are tapped from the bottom of the furnace while zinc leaves the shaft as a vapour in the carbon monoxide carbon dioxide furnace gas. The vapour is shock cooled in a lead splash condenser and is absorbed by the molten lead. Next the lead, containing zinc in solution, is pumped out of the condenser and rapidly cooled, causing the zinc to be rejected from the solution and float on the lead. The zinc layer is poured off and cast, while the lead is recycled. One surprising feature of this process is that zinc is deliberately dissolved in lead and then separated from it to yield a zinc containing about 1% lead. To obtain the high grade zinc required by the die casting industry the impure zinc can be fractionally distilled in silicon carbide columns. There are now 13 Imperial Smelting furnaces, throughout the world, capable of producing a total of about 1 million tonnes of zinc per annum.

The development of zinc die casting alloys

In the past, many hundred zinc alloys have been tested to assess their suitability for die casting, and new alloys are still being developed; some are mentioned on page 107 but so far none of them have displaced the alloy compositions which have been preferred during the past 50 years. The alloy specified in the UK as BS 1004A and in the USA as ASTM B240 is widely employed. A similar composition, BS 1004B is also specified, but is little used in the UK. Both alloys are based on zinc of greater than 99.99% purity, plus about 4% aluminium and 0.04% magnesium. The second alloy contains, in addition, 0.75–1.25% of copper. The composition of these alloys is shown in *Table 9.1* and the specification numbers of other countries are summarized in *Table 9.2.* on page 83.

A draft ISO specification under consideration in the early 1980s refers to a similar alloy with 3% copper, which finds only few applications in pressure die casting now. Some major producers in France, Germany, Scandinavia and Switzerland use the equiva-lent of BS 1004B because they find that it casts even better than the copper free alloy.

TABLE 9.1 Chemical composition of zinc alloy BS 1004 - 1972

	ALLOY A		ALLOY B	
	Ingot (%)	Castings (%)	Ingot (%)	Castings (%)
Composition				
Aluminium	3.9–4.3	3.8–4.3	3.9–4.3	3.8–4.3
Copper	0.03 max	0.10 max	0.75–1.25	0.75–1.25
Magnesium	0.04–0.06	0.03–0.06	0.04–0.06	0.3–0.06
Impurities*				
Iron	0.05	0.10	0.05	0.10
Lead	0.003	0.005	0.003	0.005
Cadmium	0.003	0.005	0.003	0.005
Tin	0.001	0.002	0.001	0.002
Thallium	0.001	0.001	0.001	0.001
Indium	0.0005	0.0005	0.0005	0.0005
Nickel	0.001	0.020†	0.001	0.020
Remainder				
Zinc				

* Single figures in this table are maxima
† An amendment of BS1004 1972 published on April 30th 1979 confirmed that Alloy A castings are acceptable with up to 0.020% nickel instead of the previous limit of 0.006%

Furthermore, it is harder and stronger. Recently die casters in Germany have adopted a lower copper content of 0.6%, and this composition is included in the German specification DIN1743, April 1978.

Zinc alloys, known as Mazak, Zamak or by other trade names, usually containing the letter Z, are unique among the commercial non-ferrous casting alloys in their need for the utmost purity and rigid control of composition at all stages from ingot to finished product in order to avoid intergranular corrosion and thus to ensure the retention of good mechanical properties.

The gradual understanding of the need for such precise regulation of composition represents a miniature history of the development of zinc die casting. The die casters of 70 years ago had gained experience and confidence in the process, using low melting point alloys based on lead and tin, with added antimony to improve hardness. It was foreseen that alloys of greater strength would widen the scope of the industry and, in about 1907, zinc alloys were used.

The pure metal was not suitable for die casting because, when molten it dissolves ferrous metals, attacking cast iron or steel dies and melting pots, causing a zinc–iron sludge to be formed. Additions of 0.5% aluminium inhibited the tendency for sludging, so that this element has been present in all zinc die casting alloys. During the 1910s, 4–8% tin was also included to promote greater fluidity. Up to 3% copper increased the strength and hardness, and for nearly 20 years, these zinc–tin–copper–aluminium alloys were used for decorative castings, components of gramophones and for carburettors. As they contained only 0.5% aluminium they could be soldered and thus repaired, or attached to other metal components.

In order to improve the strength of the zinc alloys even further, compositions with up to 12% aluminium and up to 3% copper were developed and proved equally castable

TABLE 9.2 Zinc alloy international specifications

	Ingots	Castings	Copper-free	Copper-bearing
International Standard ISO	R301		ZnAl4	ZnAl4Cul
Australia SAA	H 63	H 64	Alloy A	Alloy B
Canada CSA	HZ.3	HZ.11	AG40	AC41
Czechoslovakia CSN	42 3560	42 3560	ZnAl4	ZnAl4Cul
Denmark DS	DS 3013	DS 3013	ZnAl4 (7020)	ZnAl4Cul (7030)
Finland SFS	3090	3090	ZnAl4	ZnAl4Cul
France AFNOR	A55−102	A55−010	Z−A 4G	Z−A4U1G
E.Germany (DR) T GL	0−1743	0−1743	Z400	Z410
India ISI	IS713	M	Alloy 1	Alloy 2
Israel SII	79/2	79/2	Zamak 3	Zamak 5
Italy UNI	3717	3718	G−Zn Al4 (ZA4)	G−ZnAl4Cul (ZA4C1)
Japan JIS	H2201	H5301	Class 2	Class 1
Norway NS	16920 16930	16920 16930	ZnAl4	ZnAl4Cul
Poland PN	H−87 102	H−87 101	ZnAl4 (Z40)	ZnAl4Cul (Z41)
Portugal NP	1−1536 1−1537		ZnAl4Mg	F−ZnAl4CulMg
South Africa SABS	25	26	ZnAl4	ZnAl4Cul
Spain UNE	37302	37306	F ZnAl4 ZnAl4	F ZnAlCu4−1 ZnAlCu4−1
Sweden SIS	14 70 20 14 70 30	14 70 20 14 70 30	ZnAl4	ZnAl4Cul
U.K. BSI	BS 1004	BS 1004	Alloy A	Alloy B
U.S.A. ASTM	B240	B86	AG 40A (XXIII)	AC 41A (XXV)
U.S. Fed. Spec SAE	J468b	QQ−Z363a	AG40A 903	AC 41A 925
W. Germany DIN	1743	1743	GD−ZnAl4	GD−ZnAl4Cul
Yugoslavia JUS	CEL 050	CEL 050	ZnAl4	ZnAl4Cul

and much stronger than the alloys containing tin. Since the industry was not yet aware of the need for strict metallurgical control, it became customary for die casters having difficulty in running a casting to add tin or lead to improve the fluidity of the metal.

The early thermal methods of extracting zinc from its ores yielded a metal of little more than 98% purity, so lead and cadmium, whose compounds occur with zinc ores, were bound to be included in the smelted zinc, while tin became involved owing to the unfortunate 'dosing' habit that existed in die casting foundries in those days.

It was not appreciated that the zinc–aluminium system was liable to intergranular corrosion when these impurities were present. The results were particularly serious because valuable contracts had been obtained from the rapidly developing gramophone industry. The die castings began to deteriorate after only a few months' service, distorted, cracked and sometimes disintegrated, causing a setback of such magnitude that for some years afterwards, potential users of zinc die castings were suspicious or downright condemnatory.

Intergranular corrosion is initiated by inhomogeneities in a metal, while electro-chemical reaction is encouraged by the presence of two dissimilar constituents, causing one of them to be attacked preferentially. It is difficult to detect the early stages of degredation with optical microscopes, although the modern development of the electron microscope and other methods of obtaining high resolutions now make it possible to observe the formation of intergranular corrosion.

During the early 1920s H.E. Brauer and W.M. Peirce[4], of the New Jersey Zinc Company, conducted a research into the impurities in the zinc alloys which opened the way for overcoming the problems. To quote from their 1923 paper, 'It is the belief of the writers that zinc–aluminium–copper alloys free from appreciable amounts of detrimental impurities will withstand long exposure to any natural atmosphere without appreciable loss of strength'. Their work was followed by worldwide research until the problems were identified and solved. It was discovered that an addition of magnesium of the order of 0.1% reduced the tendency to intergranular corrosion but, much more important, new refining processes produced zinc of greater than 99.99% purity, thus removing the causes of the deterioration.

The lessons learned from the research programmes halted the practice of dosing the die cast metals with tin and lead. Furthermore it established that the optimum aluminium content was 4% – a composition near to the eutectic, as will be seen from the equilibrium diagram on page 86.   Having achieved a 'super pure' quality of the zinc, the magnesium content could be reduced to 0.04%, thus preventing the tendency to hot shortness, caused by the greater amounts of magnesium (about 0.08%), which had been necessary when the zinc was less pure.

At that time, the alloy containing 3% of copper was favoured but, when careful measurements were made, it was found that a slight dimensional change occurred with the 4% aluminium 3% copper alloy and that a reduction in copper content led to greater stability. The problems of the early die casters and the researches that led to those problems being overcome were well summarized by H.L. Evans[5] in 1937.

These developments made it possible to use zinc alloy die castings for the manu-facture of stressed components, and they could be bent, swaged and rivetted. Further-more the new die castings did not deteriorate with age though they undergo a very slight shrinkage and some increase in hardness. By the second world war, zinc die castings played an important part in the rapid production of shell and grenade fuses, gun sights, instruments and many components of military transport. Today the mechanical strength and good casting characteristics of the alloys have resulted in the wide use of zinc alloy die castings in most branches of manufacturing industry.

The compositions of the two main alloys are shown in the *Table 9.1*. Specification standards lay down stringent limits for the amounts of tin, lead, cadmium, indium and thallium in the alloys, since all those metals are harmful and, in one way or another, they are liable to be present either in the ore or in the die casting plant. Other elements such as iron or nickel must be controlled to wider limits though, compared with the specifications for other non-ferrous alloys, even these limits are stringent.

References

1.   DOWSING, R.J.
      Zinc: from protective coatings to formed superplastics. *Metals and Materials*, **10**, 19 (March 1976)

2.   BURKIN, A.R. (Editor)
      Topics in non–ferrous extraction metallurgy. pp. 104–130. Blackwells Scientific Publications. Oxford (1980)

3.   COCKS, E.J.; WALTERS, B.
      A history of the zinc smelting industry in Britain. pp. 4–9. Harrup, London (1968)

4.   BRAUER, H.E.; PEIRCE, W.M.
      The effect of impurities on the oxidation and swelling of zinc–aluminum alloys. *Transactions of the American Institute of Mining and Metallurgy* Vol. 68 pp. 796-832 (1923)

5.   EVANS, H.L.
      Zinc–base diecasting alloys. *Metal Industry*, **51/5**, 105–109 (July 30th 1937); **51/6**, 139 (August 6th 1937)

Chapter 10

The metallurgy of zinc alloys

Zinc and aluminium form a binary eutectic with a melting point of 382°C at 5.0% aluminium, a composition giving maximum fluidity but rather low impact strength. Alloys containing less than 3.8% aluminium have poor castability and inferior tensile strength. Impact strength is at a maximum in the 3.4% range but is sharply reduced when more than 4.5% aluminium is present. Consequently an aluminium content at about 4% gives the best combination of fluidity and mechanical properties.

The equilibrium diagram[1] of the zinc−rich zinc−aluminium alloys is shown in *Figure 10.1*. A typical 4% alloy has a melting point of 387°C and a short freezing range of 5°C; it has a fine grained structure, leading to optimum tensile strength, hardness and impact resistance. On solidification at 382°C primary crystals of the zinc-rich α phase, having a close packed hexagonal lattice structure, are embedded in a matrix of the lamellar

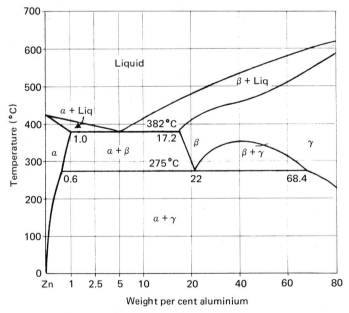

Figure 10.1 Zinc−aluminium equilibrium diagram

eutectic which is a mixture of this phase and a β phase (face centered cubic) containing 17.2% aluminium. As this alloy cools to 275°C, aluminium is precipitated from the α phase since its solid solubility in zinc decreases with reducing temperature. The primary zinc-rich phase which contained 1.0% aluminium at 382°C now contains only 0.6% aluminium at 275°C while the β phase has a composition of 78% zinc, 22% aluminium.

Table 10.1 shows the solid solubility of aluminium in zinc from the eutectic at 382°C to room temperature.

TABLE 10.1

Temperature (°C)	% Aluminium by weight
382	1.0
350	0.88
300	0.70
250	0.50
200	0.32
150	0.18
100	0.10
Room temperature	0.05

On cooling through the temperature of 275°C the alloy undergoes a eutectoid decomposition into the α phase with 99.4% zinc, 0.6% aluminium and a γ phase with 31.6% zinc and 68.4% aluminium. This aluminium-rich phase cools to room temperature, retaining some excess zinc in solution whilst the zinc-rich α phase continues to precipitate aluminium from solid solution *Figure 10.2*. Owing to the rapid cooling of pressure die castings, a metastable condition exists where the various structural changes, occurring independ-

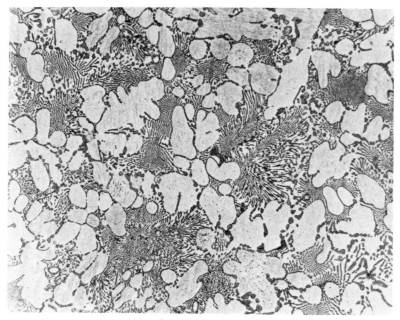

Figure 10.2 Photomicrograph of zinc alloy A showing dendrites of zinc containing a small amount of aluminium embedded in a eutectic of zinc and aluminium

ently of each other, are completed over a period of time, the major part within a month and the remainder over a period of years.

The first change to occur is the decomposition of the supercooled β phase into a eutectoid of α and γ. This transformation is accompanied by a shrinkage amounting to about 0.06%, but with no appreciable change in mechanical properties. Next, super saturated aluminium in the form of γ phase is precipitated from the zinc-rich α phase. This is accompanied by a volume contraction of about 0.06% within 5 weeks, a fall in tensile strength and an increase in tensile elongation.

The effect of copper additions on the zinc—aluminium alloys

During the 1930s, when the metallurgical problems of the alloys were beginning to be understood and overcome, the alloy with about 3% copper was favoured. Not only did it provide good strength and castability but the copper was found to combat intergranular corrosion. However, when the full benefit of using 'four nines' zinc was realized, it became unnecessary to rely on such a high copper content, and the greater stability offered by the alloys with no more than 1.25% copper was preferred. Similarly the amount of magnesium could be reduced.

The ternary eutectic point occurs at 375°C at a concentration of 7% aluminium and 3.9% copper[2]. The maximum solubility at the eutectic temperature is 1.3% aluminium and 2.9% copper. This decreases to about 0.9% aluminium and 1.9% copper at 274°C. When the zinc—aluminium—copper alloys were studied by A. Burkhardt[3] in 1936, he gave the probable room temperature solid solubilities in the zinc-rich α solid solution as 0.1—0.3% aluminium and 0.6—0.8% copper. Three phases can occur in the zinc—aluminium— copper alloys. The α and β phases correspond to those in the zinc—aluminium binary system.

An ϵ phase is a copper-rich solid solution with a hexagonal atomic lattic corresponding to the ϵ phase in the binary copper—zinc system. The copper-containing zinc base alloys contain all three phases at temperatures immediately below the solidus. On cooling, the β phase decomposes as in the binary zinc—aluminium system. Copper exerts a restraining influence on the rate of β transformation. In the ternary alloys, the solubility of copper in the α phase decreases with the fall in temperature and under equilibrium conditions the copper-rich ϵ phase is precipitated. However, in the rapid cooling obtained in die casting this is suppressed so that a metastable condition results. In the alloy with 3% copper, precipitation continues at room temperature over a period of years until stable conditions are reached. The change can be accelerated considerably by increase of temperature.

Magnesium

Magnesium additions restrain the β transformation in the zinc—aluminium alloys. At the required addition levels, magnesium inhibits intergranular corrosion and it suppresses the effect which traces of impurities have on subsurface corrosion. Nowadays when the use of very high purity zinc is universal, the magnesium content is specified at 0.03—0.06%. Higher magnesium contents lead to hot shortness and decreased impact strength. Nevertheless, it is still necessary that the specified amount is present. Any loss of magnesium through overheating or inefficient fluxing leads to a pronounced deterioration of metal quality. The magnesium is present in the form of a ternary zinc–aluminium–magnesium eutectic[4]. The lower melting point of this eutectic accounts for the hot shortness caused by magnesium contents above the currently specified levels.

Iron

At the melting point of zinc alloy, the solid solubility of iron is only about 0.001%. Unalloyed zinc shows a strong tendency to combine with iron but, fortunately, this effect is much reduced in zinc alloy owing to the inhibiting effect of the 4% aluminium content. Iron tends to combine with the aluminium, forming an intermetallic compound $FeAl_3$, causing poor castability, hard spots and difficulty in plating.

Under good working conditions, iron content can be kept below 0.02%, but if the metal is overheated in the crucible the compound $FeAl_3$ is formed. When zinc alloy is melted in cast iron pots it is necessary to make regular observations of the surface condition of the cast iron. After about 6 months continued use, the pot develops crazing, providing conditions under which iron pick-up takes place; this signals that the pot must soon be taken out of commission, or treated to reduce attack of the zinc alloy.

The effect of impurities

Lead has no detectable solid solubility in pure zinc and, under microscopic examination, even amounts of only 0.001%, can be seen as a separate phase, at grain boundaries and at the interfaces of dendrites. When aluminium is present in the die cast alloy the harmful presence of lead in such small quantities is harder to detect but can be seen under the high magnification of the electron microscope.

Tin is not likely to occur in significant amounts in zinc of 'four nines' purity but the danger arises from tin-containing materials such as solder. It forms a eutectic at 198°C with zinc (91% Sn) and this separates at grain boundaries. Consequently in addition to its effect on intergranular corrosion, the presence of tin leads to hot shortness and reduced impact strength. Because its effect is even more harmful than lead, tin is specified to be at less than 0.001% in ingots and 0.002% in die castings.

Cadmium is detrimental in amounts above 0.01%; it causes hot shortness and reduced resistance to corrosion. The minute amount normally present in 'four nines' zinc has no ill effect and the current specifications with cadmium not more than 0.005% have, therefore, been set to exclude contamination from outside sources. No intermetallic compounds are formed and cadmium has a negligible solid solubility in zinc.

The solubilities of iron, nickel, chromium, manganese and silicon in zinc alloy with 4% aluminium were studied by Friebel, Lantz and Roe[5] and the following information about impurities is taken from their 1963 paper to the American Society of Metals. Nickel is soluble in molten zinc alloy to a very limited extent and its solid solubility is only about 0.005%. It is specified as under 0.02% in castings in both Alloys A and B. Chromium also has a negligible solid solubility, and it exists in zinc–aluminium alloy as the intermetallic compound $CrAl_4$.

The solid solubility of manganese is less than 0.001%; in the paper referred to above it is suggested that manganese may exist either as Mn_2Al_5 or possibly both MnAl and $MnAl_3$. Silicon may be introduced in very small quantities from alloying additions of aluminium with the zinc. Its solid solubility at just below the melting point of zinc alloy is 0.025% and at room temperature about 0.015%.

The presence of thallium in zinc die casting alloys promotes intercrystalline corrosion, it being as harmful as tin in this respect. Thallium is a commonly occurring impurity in electrolytically refined zinc and care is necessary when selecting zinc for alloy manufacture to avoid exceeding the 0.001% maximum limit required by current specifications.

Indium is the most injurious impurity in zinc die casting alloys; amounts as small as 0.001% can cause subsurface corrosion. The maximum level allowed by BS 1004 is only 0.0005%. Fortunately, indium is not usually present in special high grade zinc or in the alloying ingredients. Antimony, arsenic, bismuth and mercury affect the stability of zinc alloy and should not exceed 0.001%. They are not present in harmful amounts in '99.99% plus' pure zinc and are unlikely to be picked up in a die casting works, so they are omitted from specifications.

Shrinkage during normal ageing

Table 10.2 shows the very small shrinkage that occurs at a progressively diminishing rate during ageing at normal temperatures over a period of 8 years.

TABLE 10.2 Shrinkage during ageing

	Alloy A (mm/m)	Alloy B (mm/m)
Shrinkage after 5 weeks*	0.32	0.69
Shrinkage after 6 months	0.56	1.03
Shrinkage after 5 years	0.73	1.36
Shrinkage after 8 years	0.79	1.41

* The measurements were made on 150mm impact test pieces and on commercial castings, air cooled.

Although the dimensional stability of alloy A is somewhat greater than that of alloy B, the difference is insignificant for most purposes. When necessary, the shrinkage of either alloy can be accelerated by subjecting the castings to a simple annealing treatment to stabilize them. The castings are held for 3–6 hours at 100°C, for 5–10 hours at 85°C or 10–12 hours at 70°C.

TABLE 10.3 Shrinkage after stabilization

(After stabilization)	Alloy A (mm/m)	Alloy B (mm/m)
Shrinkage after 5 weeks	0.20	0.22
Shrinkage after 3 months	0.30	0.26
Shrinkage after 2 years	0.30	0.37

Table 10.3 shows the further shrinkage after the stabilization treatment and it will be seen that after a short period the dimensions of the copper-free alloy A test pieces remained constant. The alloy B containing some copper does, however, continue to experience small dimensional changes.

Figure 10.3 shows a zinc alloy die cast frame for an Oertling laboratory balance which was given the stabilizing treatment. Mirrors are mounted 200 mm apart and this dimension must remain permanently accurate to within 0.008 mm in order to obtain the required accuracy of weighing.

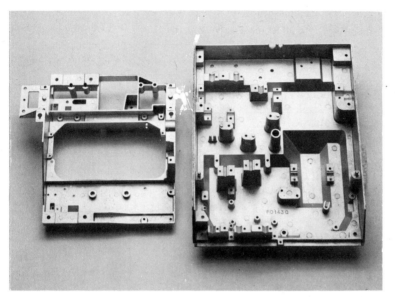

Figure 10.3 The large complex frame casting for the Oertling laboratory balance was cast to 0.008 mm in 200 mm length

The conditions of die casting manufacture entail a rapid cooling in the die and often a water quench as soon as the casting is ejected. In that connection research in Germany was reported in 1973[6] and an edited English translation was prepared by the Zinc Development Association[7]. The German tests were reported in considerable

TABLE 10.4

Die temperature (°C)	Cooling	Shrinkage after 150 days		Shrinkage after 5½ years	
		Alloy A	*Alloy B*	*Alloy A*	*Alloy B*
180	Air	0.8	0.7	0.9	1.0
180	Water	1.3	1.4	(Not reported)	

detail but *Table 10.4* shows the results of most interest here. These figures indicate a smaller effect due to the copper in alloy B, than in the earlier UK tests. However this may be because the German alloy contained only 0.8% copper and because the UK figures were for maximum shrinkage recorded in any of the samples, and not an average.

Quality mark schemes for zinc alloy die castings

To ensure that zinc alloy engineering components are manufactured under satisfactory conditions and designed correctly, codes of practice have been established in most industrialized countries. In the UK the first was published in 1955; the latest edition, 'Zinc Alloy Pressure Die Castings for Engineering' was issued as BS 5338, 1976. The code is concerned with material selection, component design, working practices and inspection procedures in compliance with the requirements of BS 1004. Not only is this standard of valuable assistance to the user of zinc alloy but it helps die casting companies to reassure enquirers that they can meet the requirements, particularly for stressed components.

Only die casting companies which hold a licence from the BSI may give evidence of their compliance with the requirements of BS 1004 by reproducing the Kite mark on their products. Castings which are too small, or otherwise unsuitable need not show the Kite mark. Once a company has been issued with a licence, renewed annually, they must comply, from a metallurgical standpoint, with the BSI requirements for supervision, inspection, testing and record keeping to ensure that the finished product will comply with the provisions of the Standard. Companies licensed by the BSI agree to undertake a daily programme of testing and inspection. In addition, BS 1004 states that the castings must be free from undue porosity; no patching, plugging or welding is permitted.

Procedures are defined for metallurgical analysis of raw materials and analysis of samples of alloy taken daily at random from machines in the plant. Testing may be undertaken at the die caster's own laboratory, approved for this purpose and subject to inspection visits. Additional requirements are concerned with routine checking of automatic temperature control equipment and the procedures to be followed when treating scrap and reclaimed metal. Material or components which are not satisfactory to the Chief Inspector must not be used in the manufacture of products covered by the licence. France, Germany, Italy, Australia and North America have quality mark schemes similar to the British one.

Metallurgical control in the zinc die casting plant

In order to produce die castings which comply with the specification, care and attention is needed throughout the handling, melting, fluxing and temperature control of melts. Pyrometers require regular checking against a standard; a prevalent fault may arise from failure to check pyrometers, leading to unsuspected overheating. The first signs of this are often shown by a warning from the Inspectorate that magnesium content in a batch of castings has dropped to dangerously near the allowed minimum.

Scrap may be divided into two groups, according to whether the metal can be remelted and used with confidence or whether, on account of thin section or oxidation, the material must be treated with caution before being circulated back to the die casting machines. An average shot produced with the new technology from a hot chamber machine comprises about 75% casting and 25% gate, runner and overflows. In such a case, the surplus material can be recycled back to the melting section without danger of the composition of the melt being affected detrimentally. However, if the shot comprised only 25% casting and 75% surplus metal, as was customary 10 years ago, 4 tonnes would have to be melted for every tonne of castings, the other 3 tonnes going back for remelting, followed again by only 25% being in the die casting. Thus there would be a danger of deterioration; for example a fall in magnesium content.

The second type of surplus metal consists of machine dribble, spillage, swarf and flash. Surface areas of such returns are high in relation to weight, so in remelting they should be plunged directly into the bath of molten alloy to increase the rate of melting and to minimize losses. Depending on the contamination or cleanness of the scrap and its proportion in the melt, fluxing may or may not be necessary. Since magnesium tends to be removed selectively by fluxes, it often becomes necessary to add it in the form of small ingot notches of magnesium–zinc alloy. Where the alloy is in regular circulation, it is usually found that magnesium needs to be added only intermittently to maintain specification levels. Due to these deviations in magnesium content and possible dispersion of intermetallic compounds (usually $FeAl_3$) in the surface layers of the melt, it is good practice to melt and treat scrap returns in a separate furnace away from the machine holding pot.

Any metallurgical control procedure must be checked constantly against the accidental addition of impurities. Lead and tin, even in extremely small quantities, are harmful to zinc alloy and are most likely to be in the vicinity of the melt in the form of solder particles, carelessly left near the machine and eventually included in flash, swarf or sweepings which are remelted. Supervisors and all personnel need to be reminded regularly that the zinc alloys have this unusual need for clinical cleanliness. Plated scrap should be treated separately; indeed, most die casting companies do not attempt to reclaim plated scrap, but sell it to a metal refiner.

The use of inserts in castings can present contamination problems when rejects are melted. Inserts containing tin and lead, bronzes, for example, or cadmium plated inserts are best avoided. As with plated castings, separate melting and treatment furnaces must be used and the final alloy composition confirmed by analysis.

Zinc alloy manufacture at the die casting plant

Many companies in the USA, and some in Britain, make their own zinc alloy, purchasing high grade zinc, usually as bundles of slabs or ingots weighing over a tonne. Pure aluminium can be added in various forms; it is not really necessary to purchase LMO ingots since pure aluminium offcuts are less expensive. Some manufacturers buy aluminium quadrants, about 10 mm square section, that have been used in electrical conductor manufacturing, and these represent an ideal form of pure aluminium. They are baled and can be dissolved in the bath of molten zinc in an immersion tube furnace such as those described on page 110 and illustrated on page 111. Other manufacturers use scrap extruded section, but that cannot be guaranteed pure aluminium and sometimes contains small amounts of added elements. In the past, companies making their own zinc alloy melted the aluminium in a separate crucible furnace and added it to the zinc, but the fact that the aluminium must be heated to over 660°C before it can be poured, introduces problems of oxidation; nowadays the immersion of solid pure aluminium into the zinc is preferred. Accurately weighed magnesium is added as small pieces, held under the surface by an inverted steel cup.

During alloy manufacture the metal is gently stirred, then fluxed and allowed to remain for about 20 min before a sample is taken for analysis. Every melt must be analysed before the alloy is used in the die casting plant, but this can be done rapidly with a direct reading spectrograph following the technique described in BS 1225 (1970). Before the development of spectrographic analysis the manufactured zinc alloy was cast into slabs, which were retained until the composition was approved; then the slabs were transferred to the bulk melting section. The incentive to conserve energy, coupled with the speed of analysis now possible, has made solidification and remelting of the zinc alloy unnecessary. A typical procedure is to have two large melter-holder furnaces, one for remelting scrap and runners and the other for alloying. As each batch of manufactured zinc alloy is approved it is pumped into the other pot, which contains melted scrap and runners that have also been analysed, so that new and remelted material is mixed. A fork lift travelling ladle then transports the molten metal to the machines.

The manufacture of zinc alloy is economic for a company having an output of 2000 tonnes or more per annum but it may be viable if a smaller manufacturer already possesses, or requires for other purposes, a direct reading spectrograph. These cost around £40 000 and they are by far the most expensive apparatus required. Apart from this, the other items include minor ancillary equipment and fume extraction.

Physical properties of zinc alloys

TABLE 10.5

Property	Alloy A	Alloy B
Melting point (°C)	387	388
Solidification point (°C)	382	379
Specific gravity	6.7	6.7
Weight (lb/in^3)	0.24	0.24
Weight (g/cc)	6.7	6.7
Thermal conductivity (cal/s/cm^3/°C at 20°C)	0.27	0.26
Thermal expansion (per °C)	2.7×10^{-5}	2.7×10^{-5}
Electric conductivity (mhos/cm^3 at 20°C)	157 000	153 000
Solidification shrinkage (inch/foot)	0.14	0.14
(mm/mm)	0.012	0.012

Mechanical properties

Although zinc alloy samples can be cast in sand or permanent moulds, the mechanical properties obtained by these processes are inferior to those obtained by the rapid cooling of pressure die castings. Figures for tensile strength and other mechanical properties are available from such publications as the Zinc Die Casting Guide, issued by the Zinc Development Association jointly with Zinc Alloy Die Casters Association[8]. These figures have been obtained from sound die cast test pieces with a gauge length of 50 mm and a diameter of 6 mm, while the impact strength figures were obtained from test specimens of 6 mm square cross-section, two test specimens 75 mm long being taken at a time from a 150 mm cast length. Figures such as those published in the above are accepted by the BSI.

Table 10.6 shows the tensile strength, elongation and impact strength at temperatures up to 95°C. Impact strengths are shown in foot-pounds and Joules, the SI units. The figures in *Table 10.6* and in *Table 10.7* show that the alloy BS 1004 B has greater tensile strength and impact strength than the copper-free alloy BS 1004 A. The dimensional change on ageing is a little more with the copper-containing alloy but the tendency to prefer the equivalent of BS 1004 B in European countries including Germany and Switzerland can be understood. Indeed a reconsideration of the possibility of changing to the alloy containing copper, especially if the copper content is set at 0.75% is perhaps overdue elsewhere.

TABLE 10.6

Temperature (°C)	Tensile strength (N/mm2)		Elongation (% on 2" (50mm))		Impact strength (ft/lbs)		(Joules)	
	Alloy A	Alloy B	Alloy A	Alloy B	A	B	A	B
20	280	346	11	8	42	46	57	60
40	247	293	16	13	42	46	57	60
95	196	239	30	23	43	–	58	–
					[Alloy B not tested at 95°C]			

Properties at subnormal temperatures

Table 10.7 below gives the mechanical properties of Alloys A and B from 20°C to − 40°C. At subnormal temperatures the tensile strength of both alloys increases slightly, but the elongation and impact strengths diminish considerably. Immediately normal temperatures are regained, the original elongation and impact values are restored.

TABLE 10.7

Temperature (°C)	Tensile strength (N/mm2)		Elongation (% on 2" (50mm))		Impact strength (ft/lbs)		(Joules)	
	A	B	A	B	A	B	A	B
20	280	346	11	8	42	46	57	60
0	296	373	9	8	8	10	51	54
−40	316	373	4.5	3	2	2.4	2.8	3.2

In 1942, the Ministry of Supply carried out tests on zinc alloy munition components down to −68°C and it was demonstrated that even at this low temperature the die castings were sufficiently shock resistant for most purposes. Indeed the impact strength of zinc alloy at normal temperatures is greater than that of cast iron, while at subnormal temperatures the impact strength is only reduced to an amount similar to that of cast iron at room temperature. It is, nevertheless, recommended that zinc alloy castings intended for use at subnormal temperatures should have adequate section thickness. The embrittlement of zinc at low temperatures is a property which is used in the cryogenic trimming of zinc alloy components to be discussed in Chaper 13.

Zinc alloy die castings in corrosive conditions

Zinc alloy die castings produced according to the requirements of BS 1004 or other relevant specifications behave in a similar manner to pure zinc[9]. They are resistant to atmospheric corrosion and that of weakly alkaline conditions but are corroded by acids. They are strongly resistant to chemical fluids such as petrol and glycol which are involved in cars and domestic appliances but less so if moisture is also present. Uncoated zinc alloy should not be allowed to come into contact with food or drinks. It is not dangerously toxic but it can taint food.

In the atmosphere, unprotected zinc alloy die castings tarnish, but the adherent coating, principally of basic zinc carbonate, is protective. In severely polluted atmospheres the rate of corrosion is unlikely to exceed 0.013 mm/year, while in unpolluted rural atmospheres the amount of corrosion may be only a tenth that rate. In confined damp places, where condensation occurs on the surface, untreated zinc alloy die castings develop a white coating similar to that seen on galvanized products. In such circumstances the chromating treatment DEF 130 is applied to protect them. When the die castings are to be installed permanently out of doors on the coast they should be chromated, followed by an epoxy powder coating. The anodizing process described on page 294 is also effective in preventing corrosion in marine conditions.

There is a wide variation in the behaviour of zinc alloy in water because of the difference in hardness and acidity in different districts. Rates of corrosion of 0.001–0.013 mm/year have been reported. Hard waters deposit a scale of calcium carbonate, which then reduces the rate of further corrosion. Some acidic waters, for example those which are peaty, cause the greatest tendency to corrosion. When immersed continuously in sea water, corrosion is of the order of 0.015–0.025 mm/year, but if the immersion is intermittent the rate of corrosion is considerably higher. Consequently any zinc die casting required to be in permanent or intermittent contact with sea water must be chromated, followed by treatment with epoxy powder coating.

Zinc alloy posesses good resistance to mildly alkaline solutions, and acquires a protective coating when in contact with cold or warm soapy water. Permanent exposure to concentrated solutions of strong alkali is not recommended, although cold dilute quiescent solutions of washing soda do not attack zinc die castings sufficiently rapidly to preclude their use in such liquids. Solutions of detergents at normal washing temperatures are also satisfactory in contact with zinc alloy; no undue corrosion has been found, although the alloy suffers slow attack from boiling solutions of detergents. Chromated zinc alloy washing machine pumps are widely used; anodized pumps have been guaranteed for 10 years in the USA.

Zinc alloy is not corroded by natural gas made from oil, as this is free from moisture, nor is it corroded by towns' gas; if a humid gas from which water may condense is used, corrosion can be prevented by chromating. During the 1960s, prejudice against zinc alloy arose because some castings of inferior manufacture containing porosity had given poor service. Then the question of suitability of zinc alloys for gas appliances was investigated thoroughly and their use was approved. Die cast zinc governors were specified in Britain under BS 3554 (1971).

Although refined oils do not attack zinc, diesel oils containing sulphur have been known to dissolve small amounts of zinc, leading to blocking of the fine capillaries of fuel jets. Lubricants, such as heavy oil or grease for use in conjunction with zinc alloy die castings, should be free from acid; mineral oil lubricants can be used with confidence.

References

1.   HANSEN, M.
      Constitution of binary alloys. 2nd edition, p. 149. McGraw Hill, New York, (1958)

2.   SMITH, C.S.
      Constitution of ternary alloys. Aluminium—copper—zinc. *Metals Handbook*, 1244 (1948).
      American Society for Metals

3.   BURKHARDT, A.
      Technologie der Zinklegierungen. Julius Springer, Berlin (1937)

4.   FINK, W.L.; WILLEY, L.A.
      Constitution of ternary alloys; Aluminum—Magnesium—Zinc. *Metals Handbook* (1948)
      American Society for Metals

5.   FRIEBEL, V.R., LANTZ, W.J.; ROE, W.P.
      Liquid solubilities of selected metals in zinc-4% aluminum. *Transactions of the American
      Society of Metals* **56,** 90 (1963)

6.   Dimensional changes in zinc die casting alloys after ageing. Material Testing department of
      Erich Herrmann & Co.Kg. *Zeitschrift für Werkstofftechnik,* **IV,** 333—340 (1973)

7.   Edited translation of Reference 6 prepared by the *Zinc Development Association,* London
      W1X 6AJ

8.   *Zinc Die Casting Guide* Zinc Development Association, London

9.   *Zinc: its corrosion resistance.* Compiled by the Battelle Memorial Institute, Columbus, Ohio.
      Jointly published by the Zinc Development Association (UK), Zinc Institute (USA) and
      Australian Zinc Development Association and others.

Chapter 11

Zinc alloy die castings

Zinc alloys can be die cast to closer limits of accuracy than alloys of higher melting points. *Figure 11.1* shows a pair of decorative castings, made in Japan, which illustrate the fine reproduction of detail which can be achieved. The comparatively low melting point also leads to die lives of at least several hundred thousand shots. Some examples exist of well made dies producing castings of even section which have lasted for 10 million shots with only minor die repairs being needed.

Hot chamber machines, such as the one shown in *Figure 11.2,* are employed for the production of zinc alloy die castings. The metal injection system is contained in the machine and the molten metal does not have to be ladled, so the development of automation was easier than with cold chamber machines. The lower casting temperature of

Figure 11.1 Two decorative models, die cast in zinc alloy. (Courtesy Furukawa Casting Co. Ltd.)

Figure 11.2 Automatic hot chamber machine with end type conveyor. (Courtesy E.M.B. Co. Ltd.)

zinc alloy was an added bonus. Automatic production of zinc die castings was being achieved in the 1950s; although very small components had been made automatically even earlier. Today most zinc alloy die castings are produced automatically, the only exceptions being components which were originally planned for manual production on old machines and with dies of such complexity that it has not been practicable to automate them. Some small companies specialize quite profitably in the execution of short run orders, often taking over dies from larger companies which no longer have the facilities for undertaking manual production.

The specialized production of very small zinc die castings is a world of its own, with automatic production rates sometimes exceeding 100 shots per minute, using machines that are exclusive to the manufacturers and not available to the wider die casting community. The quantities involved are immense; for example one British company makes 4 million umbrella tips per month while another has manufactured panel nuts with

Figure 11.3 Group of small zinc alloy die castings. (Courtesy Dynacast International Ltd.)

a built-in internal thread required for the electrical and electronic industries with an output of 50 million pieces per month. Other parts include wheels for model trains, cigarette lighter parts, small gears, locks, cams and wing nuts. *Figure 11.3* shows a group of such small die castings, ranging in weight from 3 to 20g.

In the USA, the Gries organization has been developing what is believed to be the world's fastest, most accurate and smallest die casting machine. Production rates of up to 50 000 per 8-hour shift have been attained; the machine is designed to make very small parts such as a watch wedge pin which weighs less than 1/300g. A French manufacturer which produces small parts with cast-in inserts converted a standard machine shown in *Figure 11.4* to make 'jet tube' castings for carburettors from a double impression die at a rate of 700 injections, or 1400 castings, per hour. The brass tube inserts are automatically fed into the die.

Figure 11.4 Automatic machine for pressure die casting small zinc alloy components with cast-in inserts. (Courtesy Siobra-Arbois)

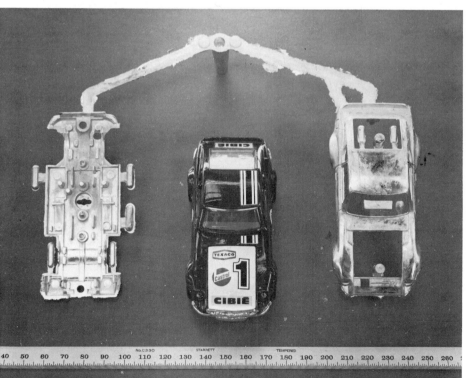

Figure 11.5 Body and base of a model car. (Courtesy Lesney Products & Co. Ltd.)

The manufacture of die cast toys and models is another specialized field. Here the achievement arises partly from clever design of the components so that they are suited for automated production, generally on machines designed by the company. Much of the technical success of these products has been due to well thought out die design, including positioning of ejectors so that the castings are pressed from the dies evenly without any tendency to stick. Such dies are made with great care to ensure that they operate with the precision of a watch and with a maximum die life.

Figure 11.5 shows the body and base of a model car, produced from a composite die. Die design took about 340 hours and pattern making about 375 hours. The double impression die required about 2900 hours to make, but, before it was put into full scale automatic production, a further 1300 hours were spent in polishing and bedding—out to ensure that all parts worked smoothly and that there were no stress raisers. By such care being exercised, followed by an adequate period of die testing, production hold ups are brought to a minimum and fast production is achieved. The die blocks were hardened to 46 HRC and the die inserts to 48 HRC. The manufacturers expect that a die life of at least 1½ million shots will be achieved before any substantial repairs are necessary.

Several toy and model making companies with a high standard of automation have extended their production to the field of trade die casting, producing small and medium sized parts required in large quantities. Their experience in 'designing for die casting', skilled toolmaking and fast automatic production has enabled them to become formidable competitors. The set of drawing instruments shown in *Figure 11.6* was produced from six die castings by the same company which made the model car illustrated in *Figure 11.5*.

The engineering developments of automation have been covered in technical papers from many parts of the world and do not come within the scope of this book. Metallurgical control and good housekeeping have not received so much publicity but are none the less vital. The die has to be made and heat treated to give trouble free production and cleaned

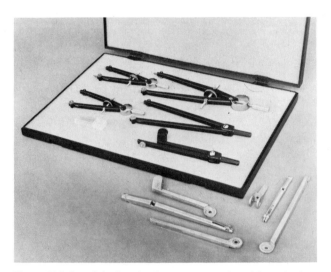

Figure 11.6 Set of six drawing instruments produced from six zinc alloy pressure die castings. (Courtesy British Thornton Ltd. and Lesney Products & Co. Ltd.)

regularly to remove flash, which would otherwise lead to castings sticking on ejection. Burnt-on lubrication deposits, which can cause surface deterioration, interfere with smooth ejection and reduce heat conduction, must also be removed. One of the most frustrating minor handicaps to full mechanization comes from ejector pins which stick when metal flash creeps between the pin and the guide. Each ejector should be identified and ground into location to minimize the production of flash.

Waterways must be cleaned frequently to ensure that the calculated rate of die cooling is maintained. A deposit on the die waterways only 0.005 in (0.12 mm) thick has been found to reduce die cooling so much that production rate is lowered by 25%. Some companies have overcome this problem by filtering the cooling water. Soft waters tend to form algae; hard waters deposit heat insulating layers on the cooling channels, so problems in one form or another must be anticipated. The use of cooling fluids, discussed on page 215 can provide an alternative method of die cooling and the same fluids can be used for die heating.

Thin-wall die casting and the new technology

Although small zinc die castings have been produced down to 0.6 mm section or less for many years, larger castings generally had wall thickness of at least 2 mm. To produce at or below this limit required even more than the usual amount of trial and error modification to the die and especially to the feed systems. Furthermore scrap rates tended to be unacceptably high. This situation has changed dramatically in recent years, thanks to extensive research programmes in the USA, Australia and the UK. The work has demonstrated the roles of many different factors in producing good zinc die castings: locking force, die temperature and the parameters of the injection cycle. Their influence on both the quality of castings and the economics of production have been established. An important outcome has been the ability to produce consistently, accurately and economically, castings much thinner than would previously have been considered practicable. Wall thickness of 1 mm or even less can now be achieved, and this is putting zinc alloy die

Figure 11.7 Zinc die cast case for industrial timer. (Courtesy Smiths Industries Ltd., and White and Edwards Ltd.)

castings into a strong competitive position. *Figure 11.7* illustrates an industrial timer, in which the upper portion of the casting is only 1 mm thick. The base portion is heavier to give stability. Such a component would not have been competitive without the 'thin wall' technique.

The essentials of this new technology are the measurement of the capability of the die casting machine to fill a die under suitable conditions, and the design of the gate and runner system so that metal flows into and fills the cavity under conditions which give a satisfactory casting for the purpose intended. These aspects have been described extensively in various publications[1,2], and what follows is a summary.

Figure 11.8 (lines P) shows the typical 'pumping' characteristic for a die casting machine, produced from measurements of pressure and flow rate. All pumps have the characteristic that as the flow rate increases, the pressure in the system drops. The position

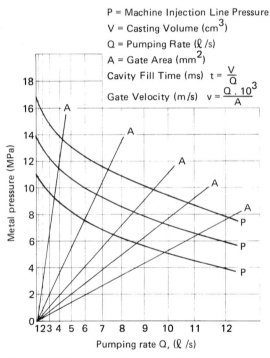

Figure 11.8 Machine performance – die design chart

and slope of the line are characteristic of the particular machine type and operating conditions. The function of the machine is to force molten metal under the required conditions into the die. Its ability to do so is limited by the hydraulic resistance of the runner, and in particular, the gate which, in a well designed system, has the smallest cross sectional area. If pressure versus flow rate is plotted for different areas, lines A, are obtained. When 'pumping' and 'resistance' graphs are overlaid, the intersection points indicate the pressure and flow rate which can be achieved by a particular machine through given gate areas. From this a range of cavity fill times and gate velocities can be predicted. Armed with these predictions, the die caster is in the best position to match production conditions to the requirements of the casting, using empirical recommendations chosen according to the form and function of the component. Matching the machine to the casting determines the required flow rate. The overall aim is to produce a runner system which is hydraulically efficient. This means that the following features should be applied. features should be applied.

1. Sharp corners and rapid changes in section should be avoided.
2. Cavitation should be reduced by accelerating the metal velocity by around 10–20% around bends.
3. Velocities should be 15–25 m/second at the nozzle, increasing this within the runner system to a gate velocity of around 40m/second.

Figure 11.9 shows the cross sectional area of the various parts of a die feeding system. Sudden changes in cross sections are not desirable and can cause separation of metal flow and associated zones of low pressure. The suggested gradual decrease in cross sectional area from nozzle to gate is indicated and is in line with published recommendations that flow path areas must reduce or remain constant from the nozzle through to the gate. An ideal runner system should also allow a short solidification time

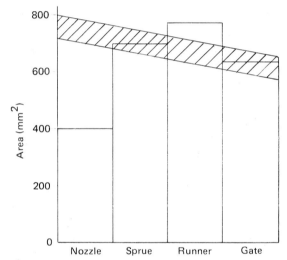

Figure 11.9 Metal flow areas of the various sectors in a die system superimposed on the suggested gradual decrease in area from nozzle to gate

for fast production rates, minimum temperature losses during injection and provide sufficient metal at the gate. Using these guidelines[3], thin-wall die castings have been produced successfully[4].

Calculating and measuring the efficiency of metal flow through goosenecks has been undertaken by the Commonwealth Scientific and Industrial Research Organisation (CSIRO) in Australia, from which the influence of gooseneck design on casting quality was identified. The main features of importance are the length and diameter of the channel and the type of bend employed. Experiments were performed on changing the size of the channel in the gooseneck and nozzle, using sand cores or cast-in tubes.

Thin-wall alloy die castings have an additional advantage beyond metal saving. The rapidly chilled outer skin of most pressure die cast parts is of the order of 0.5mm thick, so a component of wall thickness 1 mm is entirely skin and the strength benefits accordingly. A die casting of 2 mm wall thickness is half skin and half weaker internal structure; thus the strength of such a casting in comparison with its weight is less than that of a thin-wall component. Australian tests have shown that the tensile strength of a test piece increased by about 40% when the section was reduced from 4 mm to 0.6 mm.

When optimum metal flow conditions have been established, the control of heat flow becomes more important. Traditionally, excess heat has been put into the die, either through the volume of metal in the casting itself or by adding overflows whose main purpose was to supply heat to the die. This excess heat was then removed by water cooling through channels whose position and size were established by trial and error. A more scientific approach is now available; analysis of heat flow into and out of the die can be made. Work is under way to enable technicians to carry out thermal analyses with minimum effort and expense, so improving the economics of zinc die casting for many applications. This new approach to die casting technology is discussed on page 231.

In achieving maximum benefit from thin-wall die casting, extremes should be avoided; it may be technically possible to make a casting of 0.25 mm section, but such a casting might be too flexible for any practical purposes. The reply to the question, 'How thin is a thin-wall die casting?' should be 'Thick enough to do the job required and thin enough to be competitive'. In this connection account must be taken of all the factors involved, not only metal and casting costs but also finishing costs, metal losses and the need for special handling of very thin castings.

The new technology has been enthusiastically adopted by companies in many parts of the world and this has led to economies in manufacture and an opening of markets which previously had been lost to other materials such as plastics. *Figure 11.10* illustrates

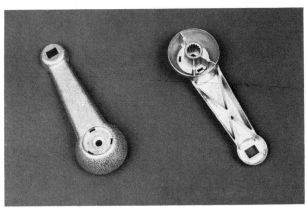

Figure 11.10 A zinc alloy window winder handle which replaced a plastic moulding. (Courtesy Ford Motor Co. Ltd.)

how the new technology enabled zinc to compete with plastic mouldings in the motor trade. The window winder handle in the Ford Fiesta and Escort had been made as a plastic moulding but proved to be too flexible. A zinc alloy die casting designed with strengthening webs shown in the illustration was more sturdy than the plastic handle. It was designed so that the original plastic knob and shear-off rivet could be fitted to the die casting. In spite of the fact that zinc is five times heavier than plastic, the design of

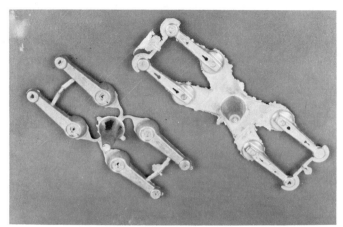

Figure 11.11 Zinc alloy die castings before and after the new technology

the die casting required a weight addition of only 20 g to the handle assembly. The die was given a textured finish to offer an attractive surface appearance on the handle. *Figure 11.11* shows how the new technology led to economies in metal. The object on the left shows the spray of castings made with the open design of sprue and the thin gates and runners. The bars connecting each pair of castings are intended for securing the handles for clip pressing. The object on the right shows a similar spray of die castings produced by the old method with a heavy sprue and a great deal of surplus metal.

Alloy developments

A new family of zinc–aluminium alloys has been developed for a wide range of engineering applications. Normally cast by the permanent mould process, these alloys are stronger than many cast irons, harder than bronze or aluminium and are challenging conventional casting materials for heavy duty applications. The original alloy ILZRO 12, developed for sand, plaster and permanent mould casting, was introduced in the 1960s by the International Lead and Zinc Research Organisation. It is a hyper eutectic zinc alloy containing about 12% aluminium, 0.75% copper and 0.02% magnesium and its interest to the pressure die caster was in prototype work, since its properties are little affected by cooling rate. Recently the aluminium content has been reduced to about 11%. Castings produced by sand or permanent mould casting techniques in this alloy have mechanical properties comparable to pressure die cast zinc alloy, enabling tests and design modifications to be completed before making the high investment needed to manufacture a pressure die. Repeated melting does not cause deterioration in casting properties. This is particularly valuable in prototype work where castings are remelted repeatedly[5].

Other zinc alloys developed in the same series for special applications included ILZRO 14 which contained titanium and possessed exceptional resistance to creep. ILZRO 16, developed from ILZRO 14, which has been withdrawn, exhibits, in addition to improved creep resistance, an outstanding dimensional stability. The nominal composition limits are titanium 0.15–0.25%, chromium 0.10–0.20%, copper 1.0–1.5% and aluminium

0.01–0.04%. The improved properties of these alloys result from the presence of dispersed hard intermetallic particles.

A group of zinc aluminium alloys[6] was developed at the Noranda Research Center in Canada, and is marketed by several suppliers around the world. Alloy ZA-12 contains 11% aluminium and is a modification of ILZRO-12. ZA-8 contains 8% aluminium. ZA27 contains 27% aluminium, 2.2% copper and 0.015% magnesium and is the lightest of the zinc–aluminium alloys. Its density is about 17% less than that of ILZRO 12, 43% less than bronze and 30% less than cast iron. Having a wide freezing range, over 94°C, it can be difficult to cast in heavier sections due to shrinkage but the alloy is suitable for applications requiring high strength and light weight especially when wall sections can be reduced to about 2.5 mm.

All of these specially developed alloys possess excellent machinability and, whilst their corrosion resistance is similar to the zinc pressure die cast alloys, they will accept protective finishes such as plating, chromating, painting and anodizing.

In contrast to zinc die casting alloys containing 4% aluminium, where melting and holding in iron crucibles is permitted, the ILZRO alloys and the ZA group tend to dissolve ferrous materials and even small amounts of iron pick-up can cause serious deterioration of casting quality. Clay graphite or silicon carbide crucibles lightly coated with refractory wash are therefore preferred. Whilst the alloys were developed for sand and permanent mould casting techniques to compete with cast iron and bronze, they can also be pressure die cast on cold chamber machines. Potential applications for these high strength alloys include castings now made in iron and requiring extensive machining, components previously made in aluminium and receiving a hard anodized finish, and bearing parts commonly cast in bronze. As a sign of their acceptance American Metal Market commenced to publish the prices of these alloys from December 1980 onwards. In 1981 the International Lead and Zinc Research Organisation published a report[7] on the properties of zinc alloys, including the ILZRO compositions discussed above.

References

1.   NEVISON, D.C.H.
      Thin-wall zinc — a new technology for die casters. *Precision Metal,* **32,** 27–31 (March 1974)

2.   RANDOLPH, J., HAYMER, P.F., THOMPSON, T.W.; REICHARD, D.
      Thin-wall zinc *Die Casting Engineer,* **21,** (4), 16 (July–August 1977)

3.   GROENVELD, T.P.; KAISER, W.D.
      Effects of metal velocity and die temperature on metal-flow distance and casting quality. *Die Casting Engineer,* **23,** (5), 44–47 (September–October 1981)

4.   DAVIS, A.J., SIAUW, H.; PAYNE, G.N.
      The significance of metal pressure in hot chamber die casting of zinc. *Society of Die Casting Engineers.* Congress Papers G-T77-072, 073 and 074 (1977)

5.   *Prototypes for zinc alloy die casting.* Zinc Alloy Die Casters Association, London (1972)

6.   DREGER, D.R.
      Zinc goes to work in the foundry. *Machine Design,* **52,** (15), 104 (June 1980)

7.   *Engineering properties of zinc alloys.* International Lead and Zinc Research Organisation, New York (1981)

Chapter 12

Melting processes for zinc alloys

Owing to the low melting point of zinc alloys and the necessity to avoid super-heating, delivery of molten metal to the die casting plant is only practicable where a smelter is adjacent. Consequently most zinc alloy die casting companies purchase ingots. In some cases 'Jumbo' ingots or slabs of pure zinc are received. In Britain, these ingots weigh about 1 tonne but some companies in Europe and elsewhere receive 2 tonne ingots. When slabs are purchased they are usually consigned in 1 tonne packs. The alloy is made up at the die casting works, as described on page 93. The zinc ingots have indentations to facilitate lifting by cranes fitted with grabs, or by fork lift trucks.

At the other extreme, in some automated plants where very small die castings are made, an alloy ingot of special shape is attached to a device above the holding crucible and is lowered into the bath of molten metal at a rate determined by the output of the automatic machine. Sometimes these ingots are produced with a square hole in one end (used for lowering into the furnace) and a projecting lug at the other (used for locating with the hole of an adjoining ingot, for stacking). However, since ingots are sometimes stacked outdoors and moisture tends to collect in the hole there is a recent tendency to omit this feature.

Other companies receive zinc alloy ingots; the weight depends on the wishes of suppliers and die casting company, but an average weight is 14 kg. A number of such ingots are stacked on to four base castings, weighing about 40 kg, with feet, so that a pack of 1 tonne ingots can be lifted by fork lift truck. The alloy is melted in a central unit and distributed to the hot chamber machines, either by fork lift truck or launder system. During the past 20 years, while the automation of hot chamber machines has been developing, the methods of metal melting and distribution have been extended and improved.

Central melting furnaces

The amount of heat required to bring zinc alloy to casting temperature is only about 68 000 kcal/tonne, compared with about 250 000 kcal for aluminium. The design and melting capability of furnaces depends on the production pattern of the plant. Medium sized concerns, manufacturing die castings for a variety of customers whose schedules are ever changing and who cannot be certain to maintain continuous three shift production. often choose to bulk melt in open topped oil or gas fired crucibles of cast iron. A typical composition is 3.3% carbon, 1.75% silicon, 0.5% manganese, 0.2% phosphorus and under 0.1% sulphur. They have capacities up to about 10 tonnes, and are flanged and mounted in the furnace allowing for a generous combustion space beneath. The burner is fired tangentially so that direct impingement on the crucible is avoided. The inside of the crucible is coated with refractory and, with careful attention, it lasts about a year with continuous or double shift working. The molten zinc alloy is removed either by a pump lowered into the metal bath or by an air motor-operated pump provided as part of the equipment.

Die casting manufacturers who can be certain of large-scale continuous production find that immersion tube furnaces are ideal. They can be fired with town or natural gas,

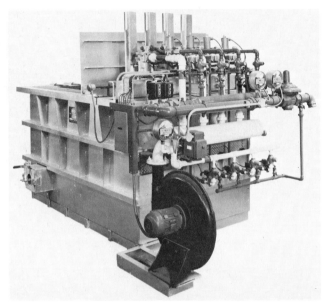

Figure 12.1 Immersion tube furnace for melting zinc alloy. (Courtesy Morganite Thermal Designs Ltd.)

propane or butane. A medium-sized British furnace is illustrated in *Figure 12.1* and a furnace of American design in *Figure 12.2*.

The steel furnace shell is lined with a high grade alumina refractory to form a bath for the metal, into which are suspended L-shaped gas fired immersion heater tubes, constructed of stainless steel and coated with refractory material. Each tube is fitted with a low air pressure nozzle mix burner mounted on the top of the vertical leg and firing down the tube. The products of combustion pass through the immersion tubes so that the alloy is not in contact with the furnace gases. Ingot and scrap are charged into the bath, where they are melted by the heat transmitted from the tubes. Molten metal can be pumped or tapped from the discharge well of the furnace, partly shown on the top left hand side in *Figure 12.1*. Such a furnace can also be supplied suitable for use with a launder distribution system. A stainless steel weir is fitted to prevent the tubes from flux attack. These furnaces have automatic temperature control; a second thermocouple is preset to switch off the burner to prevent overheating of the tubes.

Figure 12.2 Immersion tube furnace with motorized stirrer and metal pump. (Courtesy M.P.H. Industries Inc.)

The range of immersion tube furnaces provides melting capabilities from 450 to over 5000 kg/hour. The number of tubes is related to the output required; for example, a 450 kg/hour furnace has two tubes, whilst a 5400 kg furnace has 12. The furnaces are intended for continuous working and normally the metal level should be kept high and not fall below about 30 cm. During weekends the temperature is lowered and the metal retained in a pasty condition; the temperature is raised in time for the commencement of production. So long as the bath contains molten metal, solid ingots can be fed and melted easily, but the furnaces are not generally used for the continual melting of solid charges. However, after an annual shut down, when the lining has been repaired and equipment overhauled after emptying the furnaces, they can be refired by stacking ingots around the tubes and turning the heat on.

Although they are normally used for bulk melting, immersion tube furnaces can be operated as individual holding furnaces in situations where large tonnages are die cast on individual machines. Occasionally they have been designed to combine bulk melting and holding. Immersion tube furnaces provide a high melting efficiency, low metal losses and good melt quality. Like all melting equipment their operation has to be u..derstood to obtain the most efficient use and they must be operated with due care and attention. For example, the tubes should be protected against possible damage by ingots deposited in the furnace; the melting efficiency should be checked regularly by monitoring the gas consumption and measuring the temperature of the exhaust gases.

Other bulk melting furnaces

Basin tilter and immersed crucible furnaces, described in connection with aluminium melting on pages 58 and 56 are used for zinc alloy melting; electric resistance basin tilter furnaces with iron crucibles are also satisfactory. Reverberatory furnaces are not suitable because the metal being melted is in direct contact with the flame which causes oxidation and an unacceptably high metal loss, particularly when thin-wall zinc alloy scrap is being melted. Rotary reverberatory furnaces are even less suitable because the movement encourages the formation and entrapment of oxide.

Although coreless induction furnaces are suitable for the bulk melting of aluminium alloys, they are too costly to warrant their use for alloys of low melting point. Channel type induction furnaces, which are cheaper, are sometimes used for zinc melting, but the problems of oxide build-up in the channels is even more troublesome than when aluminium is being melted for die casting.

Some use of electricity has been made for holding zinc in the hot chamber machine. In contrast to fossil fuels, there are no products of combustion to cause stack energy losses but good insulation of the furnace walls is needed if maximum benefits are to be achieved from electricity. Sheathed resistance heaters ranging from 10 kW to 30 kW rating are sufficient for most zinc machine pots; benefits claimed are energy savings and reduced furnace maintenance. To avoid the risk of spilt metal damaging the elements, it is necessary to have a good seal around the pot.

Temperature control of the interconnecting tubular nozzle between machine holding pot and die gives extended nozzle life, reduces downtime from freezing of metal and can lead to energy savings. One system for automatic gas heating employs solenoid-controlled gas valves, while another development uses stainless steel sheathed electric resistance heating elements wrapped around the nozzle. A thermocouple is cemented into a groove on the outer face of the nozzle and the whole unit is encased with a steel shroud into which insulation material is packed. Either system of automatic nozzle temperature control ensures that heat is concentrated at the nozzle area only, and, in contrast to manual control, this will allow the fixed machine platen to be maintained at much lower temperatures. Faster cycling times can be achieved with a reduced tendency for blistering of castings.

Transport of molten zinc alloy

Metal pumps to transfer molten alloy from the bulk melting unit to a travelling ladle are practical propositions and are used almost universally. The impeller, spindle and other parts in contact with molten metal are of stainless steel, while other parts such as the pump housing are either of cast iron or chromium cast iron. Pumps of standard design are available but some die casting manufacturers either make their own or have them assembled by local contractors. With sensible maintenance the pumps last for several years without renewal of components. They can be operated by air or electrically and are usually retained suspended above or near the bulk melting crucible, then lowered into position when metal is to be pumped into the travelling ladle. The routine of transferring the metal to the hot chamber machine is similar to that for cold chamber machines, described on page 64.

Figure 12.3 Four-tube furnace connected to a launder system. (Courtesy D.K. Furnaces Ltd. and M.P.H. Industries Inc.)

Launder systems

Some large producers who can rely on a continuous demand for their zinc alloy die castings and who have well-trained metallurgical and technical supervision use launders to transfer molten alloy from the bulk melting furnaces to the machines. *Figure 12.3* shows an American designed gas fired immersion tube furnace connected to a launder, the first part being shown open and the remainder with the refractory cover. This particular equipment includes a gas fired heater tube which passes along the launder to keep the zinc alloy molten but most launders have electric elements coiled in the refractory covers or gas jets set at regular intervals.

In one British plant manufacturing die castings for automobile electrical systems, zinc alloy ingots, runners and unplated scrap are melted in a group of five basin tilting furnaces, mounted on a floor level turntable which can be indexed round so that the furnaces are moved to five positions in sequence. At the first position a nearly empty furnace is immediately below a loading chute and remains there while ingots, runners and scrap are loaded by means of a climbing conveyor. When the crucible is full, it is moved round and the second furnace is indexed into position. In the next three stages the metal is brought to melting temperature; then dross is skimmed off and a sample of the melted metal is sent to the laboratory for spectrographic analysis.

In the fifth position a pump is lowered into the crucible and molten alloy is transferred to an electric induction holding furnace, adjacent to a cast iron reservoir at the head of the launder line. Probes positioned above the metal in this reservoir signal when the metal level drops and the induction furnace tilts to replenish the reservoir. There is also a manual control available. This system ensures that the metal level in the launder is constant.

The launders are steel channels lined with refractory and with covers into which electric heating elements powered at 50 volts, are set as shown in *Figure 12.4*. The launder

Figure 12.4 Launder heating element. (Courtesy Lucas Industries Ltd.)

line is manufactured in metre lengths; replacements can be done easily and if machines have to be moved, the launder can be rearranged so that the channels leading from the launder to the machines are placed in the required positions. When ordering machines for a plant with a launder system, certain features have to be specified; for example the machines must be provided with jacks for lifting or lowering according to the runner position of the dies. Horizontal offset positions are easily accommodated.

The metal level throughout the system is the same and if one fast operating machine producing heavy castings is requiring a larger volume of molten metal than the others, it helps itself to the required amount. Continuous production is the key requirement since a breakdown of the system might delay production of several machines. During break periods gas jets are played on the junctions of the launder to each die casting machine to keep the metal molten. It is possible to isolate any machine from the molten metal supply, for example if one is being repaired, by using a small water cooled plug to chill the metal locally. If it is necessary to turn off the launder heaters, the metal must first be removed by casting out below the normal level, otherwise the shrinkage of the solidifying metal would damage the refractory lining of the launder.

Melting of scrap

In the mechanized zinc alloy production line the die casting, attached to the runner system, is usually clipped near the machine, it continues in one direction towards packing and despatch, while the runner and overflows are returned to the bulk melting unit[1]. If the automatic production system involves dropping the castings into water the returned material is either allowed to stand until all moisture evaporates or, in the case of a continuous operation, material can be passed through a drying oven.

Scrap should always be circulated as quickly as possible and if storage for a short time is necessary, it must be kept in a dry place, protected from dirt and moisture. All operators must be trained in safety procedures and warned that even a trace of moisture in a cavity of a casting to be remelted may cause an explosion if dropped into molten metal.

With the development of automatic die casting lines, particularly those with quench conveyors and automatic die cavity lubrication, there is often an appreciable amount of spillings, sludge and scrap contaminated with oil. Since the remelting of this fine and contaminated material is wasteful on flux and leads to high metal losses, it is often separated from the heavier material and sold to a refiner. When remelting scrap and runners, a typical sequence is to load the lightest scrap first into the liquid heel of metal to reduce the chances of oxidation, then the remainder is added. The temperature of the metal is raised to about 450°C and one or other of the following processes are applied.

In the one method, fluxing salts are spread on the melt and stirred using a mechanical paddle. After a few minutes the reaction is complete and the fine powdery dross, which is low in metallics, is removed with a perforated ladle. The bath is then allowed to stand for a few minutes, the surface finally skimmed clean again and the metal is ready for use. The alternative method is to by-pass the fluxing treatment; the skimmings from the surface of the melt are either sold to a refiner or, when accumulated to a sufficient quantity, are treated separately in a bulk melting furnace with fluxing salts. In both cases, metal for transfer to the die casting machines is taken from below the melt surface so that transfer of oxide and sludge will be avoided.

A magnet dragged along the bottom of bulk melting furnaces or the machine pot will quickly remove any materials such as small steel inserts accidentally returned to the melt. This will avoid damage to metal pumps, or machine goosenecks that would occur by these foreign materials.

Melting of plated zinc alloy scrap

Some die casting plants have practically no plated scrap, while others, for example those making car door handles, are bound to accumulate plating rejects containing copper, nickel and chromium. The copper is soluble in zinc alloy and is acceptable up to 0.1% in BS 1004A and up to 1.25% in 1004B. Nickel is only soluble to a limited extent and is acceptable up to 0.02% in both alloys. Chromium is insoluble, so would not dissolve in the molten metal. Therefore the danger of making out-of-specification zinc alloy die castings from plated scrap arises principally from nickel.

The following example indicates the amounts of the three elements which would be introduced by the melting of a tonne of plated scrap, such as exterior car door handles.

The plated handles are assumed to weigh 0.25 kg and to have a surface area of 50 cm². The plated finishes are taken as follows:

Copper 8μm thick, density 8.95g/cc
Nickel 25μm thick, density 8.90g/cc
Chromium 3μm thick, density 6.92g/cc

Multiplying the thickness of plate by the area, the volume for the 4000 die castings is shown below and then the weight is obtained by multiplying the volume by the density.

TABLE 12.1

Element plated		Volume (ccs)	Weight (g)	
Copper	0.0008 x 50 x 4000	160	160 x 8.95	1434
Nickel	0.0025 x 50 x 4000	500	500 x 8.90	4450
Chromium	0.0003 x 50 x 4000	60	60 x 6.92	415

Thus the amount of copper represents 0.143% of the total weight; nickel would be 0.445% — well over the allowed percentage. Chromium would be 0.04% of the total weight. Judged by the amount of nickel, it would be incorrect to use the remelted plated scrap without treatment or 'dilution'.

Some plated zinc alloy die castings that do not have to withstand outside atmospheric conditions may have a nickel plate of only 8—10 microns, so in the exercise discussed above the contamination would be much reduced. On the other hand the modern trend to produce thin-wall zinc alloy die castings means that the proportion of plate to casting is increased.

As a general rule it is recommended that if the total weight of plated scrap is under 3% it can be introduced into the remelting furnace along with unplated scrap and runners. If a greater amount of plated scrap is in circulation it is best to deal with it separately, skimming and fluxing with care, leaving the fluxed scrap for a few minutes and then reskimming. The metal is then cast into specially shaped ingot moulds, analysed and introduced into further melts a little at a time. In some plants the amount of plated scrap is too much to justify it being included in each melt but not enough to require it to be dealt with separately as described above. Such companies usually accumulate the plated scrap and sell it to a metal refiner.

References

1.   *Zinc alloy melting practice* (Advice compiled by Zinc Alloy Die Casters Association, p. 21 (December 1973). Available from Zinc Development Association, London

Chapter 13

Cryogenic trimming of zinc alloy die castings

Zinc alloys exhibit a significant change in some of their mechanical properties below −40°C and this phenomenon is used for deflashing and gate removal as an alternative to the more traditional and time consuming trimming methods. *Figure 13.1* shows the effect of decreasing temperature on the tensile strength of zinc and the impact resistance, measured by the Charpy test on un-notched bars. Consequently it is possible to embrittle zinc alloy die castings. With the additional aid of mechanical vibration or component tumbling, or both, flash, gates and overflows can be broken cleanly from the castings. Upon return to normal temperatures, the original properties are restored.

Zinc is not the only metal to show this embrittlement at subnormal temperatures[1]. It depends on the effect of temperature on the delicate balance in metals between the tendency to flow and the tendency to crack. Zinc becomes embrittled at low temperatures, whereas its sister metal, cadmium, does not. Simple metals, such as aluminium, do not embrittle, whereas transition metals, such as iron, do. Subtle variations in chemical bonding make the difference. It has sometimes been stated that embrittlement at subnormal temperatures is characteristic of metals which have body centred or close packed hexagonal crystal structures, while those with face centred cubic atomic lattices, do not show this phenomenon but too many exceptions exist for the statement to be generalized. It is more likely that the distribution of electrons in the chemical bonds of metals is the important factor.

The process of cryogenic trimming[2,3] or, as it is sometimes called, freeze barrelling, works on the refrigeration impact principle. It was originally developed for deflashing of rubber mouldings, since that material also becomes brittle at subnormal temperatures and has also been used for deflashing of plastic mouldings.

The equipment consists of a thermally insulated tank for the storage of nitrogen, sited as near to the foundry as possible. The liquid nitrogen is fed through a lagged pipe to a solenoid valve attached to a tumble barrel, illustrated in *Figure 13.2* a hexagonal tub constructed of two skins of stainless steel with about 50 mm of polyurethane foam between the outer and inner layers. In some equipment the inner skin is of stainless steel and the outer skin of mild steel. The unit is powered by a D.C. pulsed motor to facilitate speed and sequence control. Usually the freeze barrel equipment is associated with a group of automatic machines from which the zinc die castings are moved via a conveyor, and are deposited in a hopper above the freeze barrel[4].

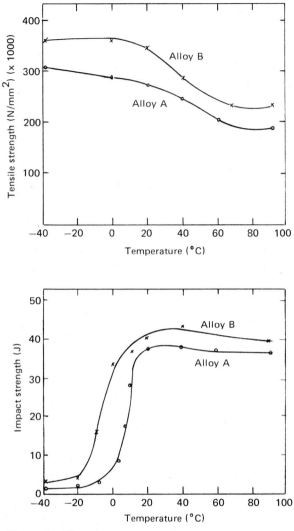

Figure 13.1 Effect of temperature on the mechanical properties
of zinc alloy

A lid closes the access aperture for processing. The barrel capacities range from
6–15 ft^3 (about 0.17–0.43 m^3). Normally the lower capacity is preferred, because the
barrel rotates on stub axles and larger barrels full of zinc die castings would cause heavy
stresses on these axles. The smaller size holds about 300 kg of sprays from the machines,
but the exact weights vary widely according to the shape and sizes of the castings and the
gates attached to them. The equipment is programmed according to the type of work
which is being trimmed and there are settings for time, speed of barrel rotation and
temperature inside the barrel.

The time sequence proceeds in three stages: first, during the pre-freeze, the barrel
rotates at about 10 r.p.m. while liquid nitrogen is continuously injected into the barrel.
When this time has elapsed the second timer takes over, the revolutions of the barrel

Figure 13.2 Freeze barreling plant. (Courtesy Brockhouse Kaye Ltd., and Air Products Ltd.)

increase to a speed which has been previously determined and set on the programme (a typical speed is 50 r.p.m.). During this second stage liquid nitrogen is injected intermittently to maintain the preset temperature; on the completion of this sequence no further nitrogen is injected and the last stage continues with the same speed of revolution.

Injection of the nitrogen is through a pipe of about 12 mm bore, leading to a number of small holes around a plate about 170 mm in diameter on one side of the barrel. A probe with a thermocouple passes through the centre of the plate to control the amount of nitrogen injected during the second stage of the process. As the injected liquid nitrogen enters the barrel, it freezes the components and then vaporizes rapidly, expanding to about 700 times its previous volume; the temperature inside the barrel is reduced to between $-80°C$ to $-140°C$, depending on the type of work being de flashed and the programme that has been set on the timer. The flash, and the join of the runner to the die casting, being thinner than the die cast components, are embrittled more rapidly. The rotation of the barrel and the presence of barrelling media assist in breaking the surplus metal from the die castings.

The deflashed castings can be discharged on to a grid or a vibratory screen which automatically separates the castings from the parts that have been detached. When the cycle is complete there are two alternative procedures. In one method a conveyor lifts the work onto a rotary sorting table and the castings are then hand picked to separate them from the waste material which is then moved to the remelt section. In the other method, the castings are discharged on to a three layer grid which is designed to separate the castings, the runners and the small pieces of removed flash.

Some users employ media ranging from carpet tacks to steel shot, but these lead to problems if by accident they become returned to the remelting furnace and are then injected into the die along with the zinc alloy. A more sensible alternative is to design the die so that additional small pieces of metal are cast as part of the shot. When the casting sprays are loaded into the barrel and rotation begins, the pieces break off and act as efficient media; when the surplus metal is returned to the remelt section, these pieces are also remelted.

A typical sequence from filling the barrel to the beginning of the next cycle is not more than 8 minutes. Under ideal conditions[5] a cycle time of 4½ minutes has been achieved: less than 1 minute for loading the barrel and 3½ minutes for the cycle of operations (30 seconds for prefreeze, 1 minute at $-80°C$ and a final 2 minutes for the last stage). The consumption of nitrogen is a vital factor in the efficiency of the process. With continuous production, and the plant near to the nitrogen container, as much as 8 tonnes of zinc die castings can be treated per tonne of nitrogen. Under average conditions 4 tonnes/tonne is still likely to achieve profitable utilization, but with broken production and temperature loss between holding tank and freeze-barrel, only 1 tonne of castings per tonne of nitrogen can be trimmed. Under these circumstances the treatment will not be profitable.

By law, the large nitrogen container must be outside the plant, and there must be ample space for the tanker to reverse to the container, (which is in an enclosed area well protected against vandalism). The freeze deburring plant inside the works, should be as near as possible to the nitrogen container. A distance of 7 m or less would be ideal, 12–15 m is acceptable but any greater distance is likely to cause wastage of nitrogen, however heavily the pipe is lagged. This becomes particularly serious if production is not continuous since every time the plant is restarted liquid nitrogen is wasted before the liquid gas arrives at the barrel. It is necessary to keep a continual check on the consumption of nitrogen, comparing it with the tonnage of castings treated.

After the cryogenic process, the castings will become frosted due to condensation of water vapour. They must not be stored in that condition or those at the bottom of a container become coated with white rust. It is therefore essential that they are dried, and not machined until they have returned to room temperature. It has been established that the low temperature treatment does not lead to any significant permanent effects on the mechanical properties of the zinc alloy[6].

For applications where cryogenic deflashing and degating can be employed, there are obvious advantages, since labour-intensive hand operation can be eliminated. In a paper given by W.G.W. Harrison[5] to the Society of Die Casting Engineers, useful information was given about the savings that can be achieved under ideal and well thought out conditions, while R.M. Jones[7] gave comparable information for cryogenic plants in a paper for the 1980 South Pacific Die Casting Congress. However, before freeze barrelling can be employed efficiently, several features must be studied to confirm process suitability. Sometimes alterations must be made to gates and metal overflow connections to ensure rapid and clean trimming.

From the economic point of view it is desirable to have as small a runner as possible consistent with casting quality because the less volume of surplus metal in the barrel the less nitrogen consumed per unit volume of finished die castings. In order to get the maximum effect from the deflashing, it may be possible to recess cores into the opposite die half, to give a vertical flash formation that can be removed by the cryogenic treatment. Die castings containing inserts require special consideration. Unplated steel inserts must be dried rapidly after treatment, to prevent corrosion; plated inserts must be given a barrelling sequence which will cause minimum damage to the surface. As with most new processes the greatest efficiency is achieved if one is prepared to adjust the product to suit the process. To get maximum efficiency in nitrogen consumption it is essential that the process is continuous with, if possible, no waiting between cycles. If for any reason the plant will be under-occupied, it is best to cut down the number of days' utilization to make sure that when the freeze-barrel is operating, it is doing so continuously.

References

1.   PALMER, J.D.
     Embrittlement is the key in cryogenics. *Canadian Chemical Processing,* **55**, (5), 56 (May 1971)
2.   TIPTON, T.T.
     Freeze deflashing zinc alloy diecastings. *Diecasting and Metal Moulding,* **6**, (4), 24 (July/August 1974)
3.   (Editorial)
     Cryogenic deflashing of zinc parts. *Die Casting Engineer,* **20**, (5), 34 (September/October 1976)
4.   CREED, Ian D.
     Freeze deburring zinc die castings. *Ninth International Pressure Diecasting Conference.* Zinc Development Association, London
5.   HARRISON, W.G.W.
     Use of cryogenics in a diecasting plant. *Die Casting Engineer,* **23**, (5), 50 (September/October 1979)
6.   *BNF Metals Technology Centre Report CE219* Investigation of effect of low temperature on zinc alloy diecasting. BNF Metals Technology Centre, Oxfordshire (June 1972)
7.   JONES, R.M.
     Cryogenic deflashing/degating of zinc diecastings. *First South Pacific Diecasting Congress* Paper No. 80 − 39, Society of Diecasting Engineers, Australia (1980)

Chapter 14

Low melting point alloys

The world's annual consumption of lead is about 4 million tonnes. The metal is obtained from three sources: the reclamation of scrap, the smelting of lead-rich ores and as a by product from the smelting of other metals. About one third is obtained from scrap battery plates, pipe sheet and cable sheathing. Impurities are removed by chemical treatment; the metal is brought to the required alloy composition, pumped from the furnace and automatically cast into ingots.

The major lead-rich ore, galena, contains the sulphide and is mined in the USSR, the USA, Australia, Canada, Peru and Mexico, these countries being named in the order of their present annual tonnages extracted. Lead sulphide is often associated with iron, zinc and small amounts of silver and gold. In the early days of lead smelting shallow ore deposits were common, and often rich enough to be smelted directly to metal. Today the remaining ore bodies are deeper, more complex and seldom rich enough to be smelted directly. The ore is ground and upgraded by flotation and then reduced to crude lead, often of sufficient purity to be used commercially; otherwise further refining processes are applied. Those lead ores which contain sufficient gold and silver are treated to extract the precious metals. In addition to the primary sources of lead, the metal is a by-product in the smelting of other metals, principally zinc; as much as 40% of all the lead that is produced comes in this way.

Nearly half of the total production of lead[1] is converted into chemical compounds, for filling battery plates, for anti-knock additions to petrol and for the paint industry. The major uses of metallic lead include components of batteries, pipe, sheet, cable sheathing and lead—tin solders. Thus the lead that is produced divides into two groups, one of which is dispersed and not reclaimable and the other which provides a large and profitable source of reclaimed metal. For example, anti-knock additives to petrol are impossible to reclaim but battery plates and lead pipes are easy to convert to good quality material. Some dissipative uses of lead, such as petrol additives, are declining in use, while the amount of lead reclaimable from batteries is bound to increase as electric vehicles replace some powered by petrol.

Lead melts at $327°C$; it is the heaviest and most corrosion-resistant of all the commonly used metals. Lead alloys have been cast since the early days of metallurgy; *Figure 14.1* shows a figurine about 70 mm high found near Luxor at the site of the ancient city of Abydos in the ruins of a temple of Osiris. It is believed that this casting was made in about 3800 BC. The Romans used lead for piping their water supply 2000

Figure 14.1 The oldest known lead casting, dating to
about 3800 BC. (Courtesy The British Museum)

years ago. Today most applications for lead and its alloys are where heavy weight and
resistance to corrosion are essential but where strength and hardness are of little significance.

Lead does not alloy readily with materials of higher strength but does alloy with
similar low strength, low melting point metals; this feature precludes the general use of
lead alloys in engineering where stresses are encountered. Some die castings are made in
pure lead containing only very small amounts of impurities but, to meet a wide range of
service requirements, various lead alloys are used. The choice of alloy depends on many
factors including density (either for weight or shielding applications), hardness, corrosion
resistance, aesthetic appearance and sound attenuation.

For centuries lead has been associated with the casting of printers' type-metal. The
low melting point of lead made it relatively easy to cast but it is soft and contracts on
solidification to such an extent that unalloyed lead type lost sharpness and definition and
would not withstand the pressure and wear of printing operations. Tin, although known
for its property of hardening lead, was not used to any great extent as it was always
comparatively expensive. Through the ingenuity of early metallurgists, antimony was
extracted from its sulphide ore and this element, in association with small amounts of tin,
conferred enhanced properties to the lead type-metal. Today the new technologies,
ranging from photocopying to computerized phototypesetting, are leading to a decline in
the tonnage of lead alloy type-metals but lead—antimony alloys, with and without other
constituents, are used for other purposes, some of which are offering increased markets.

Parts for X ray equipment and for gamma ray shielding in nuclear waste containers
are large and thick, often weighing over 3 tonnes. Generally, sand castings or permanent
mould castings are used because pressure die casting cannot achieve the required solidity.
Many components of lead-acid batteries, where the slightest evidence of porosity is
harmful, are also often made as permanent mould castings. On the other hand, some

Figure 14.2 Small lead alloy pressure die castings. (Courtesy Associated Lead Manufacturers Ltd.)

manufacturers pressure die cast the spines of heavy duty batteries. In other fields, lead alloys have a fairly wide application, in agricultural, medical, security, photographic and communications equipment. *Figure 14.2* shows some very small items for cameras, tachometers, instruments and a conical washer. Generally lead alloys are die cast by specialists.

Metallurgy of lead based alloys

The equilibrium diagram of lead—antimony alloys is shown in *Figure 14.3*. A eutectic is formed at 252°C with 11.2% antimony and 88.8% lead; it consists of alternately deposited crystals of the two metals. The solid solubility of antimony in lead is about 3.5% at the eutectic temperature but decreases to about 0.1% at room temperature. This causes age hardening by precipitation of the antimony and up to the eutectic composition antimony increases hardness and creep properties, trebles the tensile strength and decreases the density of lead by 8%. The fluidity of lead—antimony alloys decreases up to about 5% antimony content but rapidly increases again up to the eutectic composition where the alloys will give sharp reproduction of mould detail. Increasing antimony content above the eutectic composition cause some reduction in fluidity and the separation of free antimony which, being lighter than lead, can cause segregation in the melt.

Small amounts of other metals can be added to the binary lead—antimony alloys to enhance their properties. Up to 0.5% tin improves castability, and toughens and hardens the alloy, though to a lesser extent than antimony. The effect on the microstructure is to reduce the grain size and refine the eutectic. The rapid cooling of die cast metal has the effect of refining the microstructure so the need for other alloying elements in lead—antimony alloys is reduced. Small amounts of other elements are added where special properties are required, an example is the die casting of battery spines, discussed later.

Melting and casting of lead alloys is relatively simple, and, since they do not attack ferrous metals, cast iron or steel melting pots are used. Oxide formation is small at the

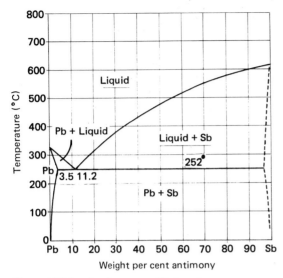

Figure 14.3 Lead-antimony equilibrium diagram

low casting temperatures and little, if any, gas is dissolved, so no special fluxing or metal treatments are required. Due to low casting temperatures and lack of affinity for ferrous materials, conventional hot chamber machines are used and high casting rates are achieved. Just as lead itself is a serious contaminant in zinc alloys, zinc is a harmful impurity in lead alloys, where it causes brittleness, intergranular weakness and dimensional instability. Not only should zinc and lead alloys be die cast in separate locations but care should be taken that not even trace amounts of zinc are introduced accidentally.

Pressure die cast automobile wheel balance weights *(Figure 14.4)*, contain between 3–6% antimony for added strength; small amounts of tin are often used to improve fluidity in the molten alloy. The integral clip of plated spring steel is cast into the weight

Figure 14.4 Pressure die cast wheel balance weights. (Courtesy Airvert Ltd.)

as an insert placed in the die before metal injection and, consequently, care must be taken to ensure that neither the plating nor temper of the steel insert is affected. Wheel balance weights are now used by all major car producers, because sophisticated front suspensions lead to any small imbalance in the wheel/tyre assembly being felt by the driver as vibrations in the steering wheel. Any such imbalance also causes increased tyre wear, due to what is known as 'tramping' in the assembly.

Lead alloy die castings in batteries

The first practical lead–acid battery was introduced in 1861, and in 1899 the world land speed record was held for a short time by Count Senatzi, who achieved 41 m.p.h. in his electric car 'La Jamais Contente', which later became the first road vehicle to exceed 60 m.p.h. Since that time the battery industry has grown in size and efficiency and is now offering an alternative to replace the declining resources of hydrocarbon fuels. Components of batteries account for the major proportion of lead consumption. The batteries are constructed by joining together the positive and negative plates for each cell, interleaving with electrically insulating separators and assembling the cells to form a battery.

Flat plate batteries[2] are widely used for storage, lighting and automobiles. A car battery is only required to start the engine and provide standby power for lights, when the engine, and therefore the alternator, is not running. Most plates for this type of battery are produced by permanent mouldings, low pressure die casting, or from rolled lead strip. Automobile flat plate batteries would have a short life if subjected to the continuous heavy duties for materials handling, mining, railways or submarines. Batteries with anti-monial lead-spined grids, located in porous lead oxide-filled, oxidation-resistant 'gauntlets' were developed to give long life in arduous conditions; in that branch of the industry pressure die castings are used successfully[3]. Felted Terylene gauntlets are placed over the whole grid or over each individual spine, the space between spines and gauntlets being filled with lead oxide. At first, low pressure die casting machines with single impression moulds were used but, in the 1960s, spined grids were made as pressure die castings, at first with single impression dies; the castings were cropped and gauntleted manually. Progress towards automation was made when a hot chamber machine was fitted with a sliding die bolster assembly carrying two dies, each with a single grid impression horizon-tally arranged. The die bolster reciprocated, so that while one grid was being cast the other was being stripped from the second die.

During the 1970s double–impression dies were made to produce about 160 grids/hour in lengths ranging from 450 to 610 mm. A 10% antimony lead alloy was used. Several mechanical problems had to be overcome before a successful production line could be operated: the ejector system had to be refined to permit closer control so that the flimsy hot grid was not distorted. Die lubrication is of critical importance in the successful production of battery grids; a water based lubricant was applied by an atomizing spray through nozzles attached to the casting extractor arm. Die temperature control was maintained by a series of electric heaters together with a water cooling system. Temper-ature indications from all die areas were transmitted to the machine control panel.

Further progress towards automation was developed, and it was decided to incor-porate cropping and gauntleting into the production cycles. A 500 tonne machine was selected to permit the operation of a four—impression die to produce grid castings 345 mm in length with 15 spines per grid. The present automated system comprises the die casting machine, the die and the handling equipment. *Figure 14.5* shows an automatic installation comprising three machines. *Figure 14.6* shows a four impression casting being extracted from the die. Each casting weighs about 0.6 kg.

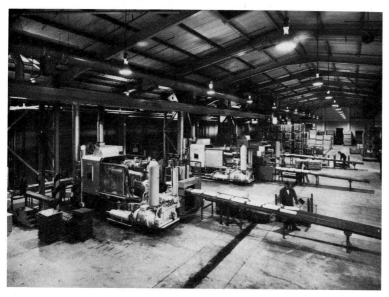

Figure 14.5 Automated production of tubular plate battery grid die castings. (Courtesy Chloride Industrial Batteries Ltd.)

Figure 14.6 Set of four battery grid die castings. (Courtesy Markham & Co. Ltd.)

A horizontal ram moves into the die space and a system of latch fingers lifts the castings off the ejectors, the ram returning from the die space to present the casting in line with a runout table. A transfer plate on this table rises, collecting the casting from the fingers and lowering it to table level. The transfer arm then moves the casting under rotary brushes which remove any flash. The arm pushes the casting to a guillotine where it is gripped and trimmed to the required length. A secondary cutter then removes the casting sprue and separates the individual grids from the spray of castings. The gauntlets are stacked in a hopper above the guillotine and individually fed on to a runout table. A transfer arm pushes the castings into the gauntlets and then pushes the gauntleted castings on to a stacking table. Currently the Chloride Group installation operates three 500 tonne Castmasters located side by side and fed with molten metal from a 3000 kg furnace by a launder system. Further developments are in hand to extend the automated systems to larger grids and to improve cycle times in the gauntleting feed mechanism.

Porosity is undesirable in battery alloys because it causes increased corrosion of the grid, thus reducing battery life and, due to the reduced cross sectional area, conductivity of the grid falls. In the pressure die casting of tubular spines, continued efforts are being made to reduce further the gas porosity levels by attention to die filling and venting. Casting of smaller diameter spines down to 2 mm are being attempted to give a weight and cost reduction and allow the introduction of batteries with increased capacity-to-size characteristics.

There have been many metallurgical developments, enabling lead-acid batteries to be introduced with extended life and reduced maintenance. Antimony, due to cost and some undesirable electrochemical effects in its use which lead to loss of water and acid mist from a battery, has gradually been reduced to 3% and less, but at these levels coarse grains and interdendritic porosity reduce the fluidity of the alloys. Low antimony levels also weaken the castings, resulting in damage during handling and finishing operations. This effect can be substantially reduced by the addition of small amounts of grain-refining elements such as tin, arsenic up to 0.5%, and copper up to 0.1%, with smaller additions of sulphur and selenium.

Table 14.1 shows the percentages of elements present in three types of lead alloy die cast spines.

TABLE 14.1

Sb	*As*	*Cu*	*Sn*	*Bi*	*Ag*	*S*	*Se*
6.3	0.3	0.007	0.11	0.01	0.005	0.003	–
3.0	0.09	0.007	0.06	0.01	0.003	0.001	0.016
3.0	0.3	0.003	0.14	0.01	0.001	0.004	–

The battery industry is highly mechanized and very competitive and it is to be expected that alternative methods of manufacture will be developed. In the field of flat plate batteries where permanent mould casting has been used, the existing methods are being challenged by expanding or punching the plates from rolled lead strip; this system is claimed to produce 200 grids/minute, which is equivalent to the output of seven permanent mould casters[4]. In addition, ageing of cast grids before assembly is not required. In 1978, 10% of the automotive batteries assembled in the USA contained expanded metal grids, and the proportion increased to 20% in 1979, although this trend has not yet been evident in other countries. Further development of expanded metal grids made from sheet lead as an alternative to cast grids will include research on lead—calcium and lead—calcium—tin alloys as alternatives to lead—antimony. Calcium has been found to offer an excellent combination of properties when alloyed with lead in low-maintenance batteries.

Up to the present time, wider use of pressure die casting in the battery industry has been limited because of the microporosity which occurs. Future developments of pore— free or vacuum die casting will probably make it possible to benefit from the high production rates attainable with automated production.

Tin alloy die castings

Tin has been used for over 5000 years; it has the lowest melting point (232°C) of all the common metals, it is corrosion-resistant, and casts with ease[5]. It is expensive, being over 20 times the price of lead, and so its major uses are limited to tin plate, where the proportion of tin to steel is less than 1%, and lead—tin solders, which so far have not been replaced as an effective way of joining metal components. Most of the world's tin comes from alluvial deposits which are worked by dredging or water-jet dispersion, while other deposits are mined underground. In each case the ore is upgraded by washing and riffling followed by flotation or other methods of separating the tin-bearing compound from the earthy material in the ore. The enriched ore is then smelted.

Tin base alloys are much more expensive than the other die cast materials, so their use is limited to two main areas. The low melting point and excellent castability of

Figure 14.7 Permanent mould cast wine jug in pewter. (Courtesy Stephan Grenningloh.)

Figure 14.8 Pressure die cast picture in pewter.
(Courtesy Anton Frieling GmbH)

tin alloys makes it possible to cast them with very close limits of accuracy; a tolerance of only ± 0.0005 mm per mm can be maintained. Numbering machine wheels have been die cast in tin alloys with even closer tolerances and, at one time, tin alloy nibs of ball point pens were die cast. Other uses include components of electrical equipment. A range of alloys is used, including those with from 8 to 18% antimony, from 1 to 6% copper and up to 15% lead. Since such parts are usually small the high raw material cost is not an overwhelming disadvantage.

Pewter[6,7], a tin-rich alloy containing 4–8% antimony and up to 2% copper, has been fabricated in the past by spinning and by permanent mould casting. *Figure 14.7* illustrates the standard of permanent mould casting attained in Germany, where a number of pewter manufacturers use that process, with an alloy containing 4.5% antimony and 0.5% copper. However, starting in Belgium and Germany in the 1970s, pressure die casting has been used effectively to manufacture tankards, candlesticks and other articles of attractive design and complex patterns. *Figure 14.8* shows a decorative pressure die casting in pewter.

References

1.   NATO Science Committee Study Group (1976)
      Rational use of potentially scarce metals. NATO Scientific Affairs Division, pp.54 and 91.

2.   MAY, G.J.
      Lead alloys for low-maintenance batteries. *The Metallurgist and Materials Technologist,* **12,** (10), 546 (May 1980)

3.   *Castmaster News* (1977–1981). Markham & Co. Ltd., Chesterfield, Middlesex

4.   MELNIK, D.C.
      The expander system. *The Battery Man,* **22,** (12), 2–6 (December 1980)

5.   *Melting, casting and working of tin and tin-rich alloys.* (16pp.) International Tin Research Institute, Perivale, Middlesex

6.   *Modern Pewter* International Tin Research Institute Publication 494, Middlesex

7.   DEAR, R.R.
      A survey of methods for producing cast pewter. Tin and its uses. Publication 102. International Tin Research Institute, Middlesex

Chapter 15

Magnesium and its alloys

The world's crust comprises about 2% magnesium, mainly as dolomite, a magnesium—calcium carbonate, and magnesite, which is magnesium carbonate. In addition, an almost limitless supply of magnesium is contained in sea water salt, and brines with a high concentration of magnesium chloride are available, sometimes as by-products of the potash industry. In 1980 world magnesium production reached a record 236 000 tonnes[1], eclipsing even the 1943 wartime total of 232 000 tonnes.

Magnesium smelting plants are located in many parts of the world where ample sources of energy are available, including the USA, Canada, the USSR, Norway, France, Italy and Yugoslavia. The metal is produced by several processes, but these can be classified as either electrolytic or thermal. As examples of electrolytic processes[2] two methods are operated at the Norsk Hydro plants at Heröya in southern Norway, both routes culminating in the electrolysis of magnesium chloride. In the older process[3], which has been operating since 1951, crushed dolomite is calcined in rotary kilns, then mixed with sea water from a nearby fjord and led to large outdoor tanks where the insoluble magnesium hydroxide settles to the bottom. The hydroxide is dehydrated and calcined, to give magnesium oxide, which is mixed with coke and mineral salts. The mixture is treated with chlorine in large shaft furnaces, producing molten magnesium chloride; this is electrolysed with a low voltage, high amperage current. Chlorine, which is recycled back to the process, is given off at carbon anodes while the liberated magnesium floats to the surface around steel cathodes.

In the newer, more energy-efficient process, brine, containing about 30% magnesium chloride, is the starting point. First, impurities such as sulphur and boron are removed and then the water is removed in several stages, first by evaporation and then by means of hot air and hydrochloric acid gas. The water-free anhydrous magnesium chloride is then taken to the electrolytic cells for conversion to magnesium and chlorine. For every tonne of magnesium about 3 tonnes of chlorine is produced and this valuable by-product is utilized in petrochemical plants.

During the 1970s the electrolytic reduction stage alone required about 20 000 kWh/ tonne of magnesium. As with aluminium smelting continuous improvements have been made to reduce the amount of energy but the power requirements for smelting magnesium will always be very considerable and will need ample sources of energy from hydroelectric stations. The process which used brine as the raw material is tending to supplant the dolomite/seawater process, partly because it requires less energy and partly because of the production of chlorine.

Several producers use thermal reduction processes in which magnesium oxide is briquetted with ferro-silicon and heated to about $1100°C$ under a high vacuum. The magnesium which is formed volatilizes and is condensed. The thermal process requires about the same amount of energy as the electrochemical process but it is associated mostly with companies handling the production of ferro-silicon. There are, of course, many advantages and disadvantages of each type of process; among the studies that have been devoted to the subject the Materials Processing Centre of the Massachusetts Institute of Technology has assessed the different processes and information has been published by the International Magnesium Association[4].

At present almost a quarter of the worlds production is from the thermal processes. The average plant size is about one-fifth that of the electrochemical plants and the capital cost is about three-quarters. Thermal plants can be expanded in comparatively small units step by step and that is advantageous in a world where costs of plant building and equipment are astronomical.

From the point of view of the die caster and the potential user of magnesium, a number of interconnected factors should be taken into account. Certainly there never need be a shortage of raw material. Massive supplies of dolomite rock occur in many countries — the West, communist countries and the Third World. In addition, the supply of sea water is practically inexhaustible: a yearly production of 100 million tonnes for a million years would reduce the magnesium content of sea water by only 0.01%. The energy requirement for the extraction of the metal is considerable although, as with aluminium, improved methods are being developed to reduce the energy per tonne of metal. A new process[4], reported in 1979, features magnesium reduction cells operating at 250 000 amp, in which production is being substantially increased with energy consumption per tonne being reduced by about 20%.

In the mid 1970s, the raw material cost of magnesium was over twice that of aluminium, but it has been estimated that by the mid or late 1980s the price will be 1.8—1.5 times that of aluminium, which, after taking the relative densities into consideration, would make magnesium viable for extended use. However, it must be noted that from the die caster's point of view the relevant relationship is not that of pure magnesium versus pure aluminium, but aluminium alloy ingots produced from secondary materials versus magnesium alloy ingots. How far this will open up many new fields for magnesium die casting depends on availability and whether the magnesium producers find other more profitable markets. About 45% of the present world tonnage of magnesium is required for alloying with aluminium, including the vast market for easy-open tops of beverage cans, made of 4.5% magnesium alloy: about 10% is used in the iron and steel industry as a desulphurizing agent and in the production of nodular cast iron. Magnesium is also used in the production of titanium, zirconium and uranium by the reduction of compounds of those metals.

The most important advantage of magnesium in comparison to other die casting alloys is its lightness. Magnesium alloys, having a specific gravity range between 1.76 and 1.83, are 30% lighter than aluminium and less than only one-third the weight of zinc and copper alloys. Other notable properties of magnesium alloys are their good machining properties, and high strength: weight ratios. The lower modulus of elasticity of magnesium compared with aluminium means that for stressed components where rigidity is important, ribs or slight increases in section thickness become necessary, but even so a weight reduction of at least 20% can be obtained. In addition to maintaining good stability under most conditions, magnesium alloys are non-sparking and non-magnetic and their 6% volume contraction is one of the lowest among cast metals, made up of 4% on solidification and 2% on freezing to room temperature.

Despite the attractive properties of this ultralight metal, there has been some reluctance to design and manufacture die castings employing magnesium alloys. This stems from magnesium's high affinity with oxygen and a tendency to connect it with fireworks, flashlight and incendiary bombs. Many of those associated with the development of magnesium have been astounded at the number of myths surrounding this metal. One example, quoted more in sorrow than in anger, relates to an Executive who learned that some magnesium components were being X-rayed along with other metals and called for the magnesium to be removed because he was confident that the metal would explode when examined by X-ray. However, once the properties of magnesium are understood it becomes evident that the metal will only burn vigorously if it is superheated after melting, or powdered, and then in contact with copious supplies of air. With proper precautions and good housekeeping a magnesium pressure die casting plant can operate as safely as one for aluminium. It is significant that those countries, such as Germany, which have a long tradition of magnesium development, use the metal on an immense scale and export their products to many parts of the world.

The scope for magnesium is considerable and, although there has been a shortage, the addition of more extraction facilities should lead to more supplies being available at a competitive price, to permit a healthy growth of magnesium die castings, particularly in the automotive industry[4]. Such parts as transmission housings, crank cases and wheels are already being used and future developments are anticipated. The problems of producing die cast aluminium inlet manifolds have been discussed on page 44 but the possible use of magnesium has been considered. For example if the inlet manifold of a large eight—

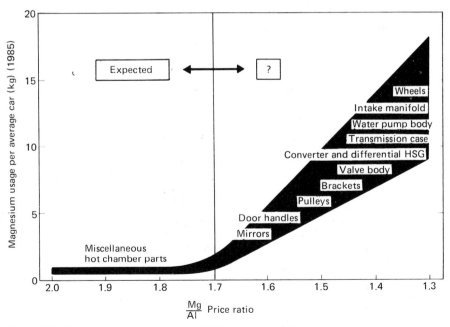

Figure 15.1 Possible magnesium usage by 1985 as a function of Mg:Al price ratio

Figure 15.2 Volkswagen transmission housing

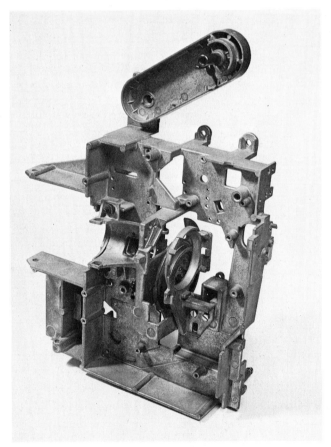

Figure 15.3 Film projector housing die cast on a hot chamber machine.
(Courtesy Eumig, Vienna)

cylinder Ford engine were converted from cast iron weighing 21 kg the magnesium die casting would weigh only 5.2 kg. Opinions on future possibilities for magnesium vary, but the Ford Motor Company have suggested that if and when the magnesium: aluminium price ratio falls to 1.5:1, a requirement of about 14 kg of magnesium per car is possible. *Figure 15.1* is a chart prepared by the Ford Motor Company which indicates the components of automobiles which might become viable for magnesium as its cost relative to that of aluminium declined. A great benefit of course will arise from the fuel savings which arise from reduced weight.

In other market sectors, chain saws, domestic tools, rotary lawn mower decks, office machines, fishing reels and instrument cases, illustrate large uses of magnesium. Many other lesser applications have been successful, including tennis rackets, archery bows and 'Little League' baseball bats. *Figures 15.2* and *15.3* illustrate some uses of magnesium alloy die castings.

The metallurgy of magnesium die casting alloys

Magnesium melts at 651°C, a temperature close to that for aluminium, but the boiling points of the two metals are very different; aluminium boils at 2057°C, magnesium at 1107°C. When molten aluminium is exposed to the atmosphere a very thin film of Al_2O_3 is formed which is such an efficient protector that the surface of the melt retains its metallic appearance. On the other hand, when magnesium is melted the vapour oxidizes to MgO which occupies less space than the metal it has replaced and the surface becomes covered with loose particles which glow if the magnesium is overheated. For this reason it is essential to encourage the formation of a protective film, and this accounts for the addition of minute amounts of beryllium, which will be discussed on page 146 and the need to hold the molten metal in an atmosphere which will prevent oxidation.

The commercial casting alloys contain up to 10.5% aluminium, with smaller amounts of zinc, manganese and some other elements. There are several alloys, without aluminium, that have been developed specially for aerospace castings where exceptional properties are required and cost is of less importance. For example an alloy containing 0.7% zirconium, 3.0% thorium and 2.5% zinc is used in American and British jet engines. However, at present, such alloys are outside the range of pressure die casting production.

In the UK, the USA and several other countries, the standard magnesium die casting alloy is based on the composition 7.5—9.5% aluminium, 0.3—1.5% zinc and 0.15—0.8% manganese. This composition is designated as BS 2970 MAG 7 in Britain, and ASTM AZ 91B in the USA. It is sometimes known by the trade name Elektron C in the UK but, world-wide, AZ 91 is the term most often used. A similar alloy, MAG 3, has a composition related to MAG 7 but is of higher purity and is used principally for permanent mould castings for the aeronautical industry.

Figure 15.4 shows the equilibrium diagram of the relevant part of the magnesium— aluminium system. A eutectic is formed at 32.3% aluminium, with a freezing point of 437°C. Magnesium holds 12.7% aluminium in solid solution at the eutectic temperature, but the solid solubility decreases to 6.2% at 300°C and 1.5% at 100°C. A photomicrograph of die cast alloy with about 9% aluminium, shown in *Figure 15.5,* consists of a eutectic structure surrounding the lighter coloured magnesium rich α solid solution. The effect of the small amounts of zinc and manganese constituents on the freezing characteristics of

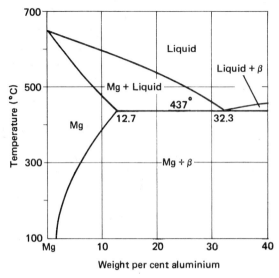

Figure 15.4 Equilibrium diagram of magnesium–aluminium alloys up to 40% aluminium

the alloy are only slight. Evidence of manganese can be seen as a manganese–aluminium intermetallic compound (manganese is virtually insoluble in solid magnesium). Since zinc is solid soluble in magnesium to about the same extent as aluminium, and as the zinc content is only about 1%, it enters into the magnesium–aluminium solid solution.

Figure 15.5 Microstructure of die cast magnesium alloy ($\times$ 500).

The compositions of the alloys principally die cast in the USA and the UK are shown in the following Table. The alloys used by Volkswagen are discussed later; in addition to AZ 81, two other alloys which they developed are included in *Table 15.1*.

TABLE 15.1 Magnesium pressure die casting alloy specifications

	UK		USA		Volkswagen	
	MAG 3	MAG 7	AZ91	AZ 81	AS 41	AS 21
Aluminium	9.0–10.5	7.5–9.5	8.0–9.5	7.0–8.5	4.0–5.0	1.9–2.5
Zinc	0.3–1.0	0.3–1.5	0.3–1.0	0.3–1.0	0.1 max	0.15–0.20
Silicon	0.3 max	0.4 max	0.1 max	0.1 max	0.4–1.0	0.7–1.2
Manganese	0.15–0.4	0.15–0.8	0.1–0.3	0.1–0.3	0.2–0.5	0.35 max
Copper	0.15 max	0.35 max	0.05 max	0.05 max	0.04 max	0.04 max

The following list shows the related specification of some other countries, and of international designations, corresponding with castings in AZ 91B.

International Standard ISO	MgAl9Zn, No. 2
France	AFNOR G–A9Zl
W. Germany	DIN 1729 MgAl9Zn1
Japan	MDC 1B
UK	BS 2970 MAG 7
USA	ASTM AZ91 and SAE AZ91B
USSR	ML6

As with most alloys, the composition specified by the different countries varies. For example MDC 1B, the Japanese equivalent of AZ 91, includes the following constituents, which may be compared with those shown in the *Table 15.1*.

TABLE 15.2

Element	Percentages
Al	8.3–9.7
Zn	0.35–1.0
Mn	0.15 min
Si	0.50 max
Cu	0.35 max

Magnesium alloy die casting at Volkswagen[5]

Volkswagen, the largest individual producer of magnesium alloy die castings, began to use magnesium in the early models of the Beetle; up to 1981 more than 20 million of these vehicles had been built, containing about 400 000 tonnes of magnesium. At first the components were made as permanent mould castings, followed by the production of small parts as pressure die castings in 1949 with further development until 1960, when the transmission housings and crankcases were pressure die cast. At the beginning of the 1970s the production of air-cooled engines came to its highest point; in 1971 more than 2 million cars were produced with a consumption of magnesium amounting to 42 000 tonnes.

In the mid 1970s new front wheel drive models were introduced with liquid cooled engines, at the same time the price of magnesium increased considerably in comparision with that of aluminium. To counteract this, and to retain flexibility, the transmission housings were specified as magnesium or aluminium die castings with alternative production lines. As a measure to decrease magnesium costs, the practice of recycling other than self-generated scrap was expanded. In the plant at Kassel the output of about 900 tonnes per month is obtained from about 20–50 tonnes of scrap from Volkswagen reconditioning and machining departments and the remainder as magnesium alloy ingots.

Magnesium die castings are made by Volkswagen at two plants in Germany, one in Brazil producing about 2000 cars per day and another in Mexico producing about 300. Although aluminium is also used, the most important application of magnesium in Germany remains the transmission housings, ranging in weight from 6.93 to 11.70kg.

The metal is melted in electric induction furnaces with steel crucibles. After melting and refining, the metal is decanted into resistance heated settling furnaces. Fork lift trucks then carry the molten material in transfer crucibles to the electric resistance furnaces at the die casting stations. Ladling into the machines is automatic, using nitrogen as a pressurizing gas. The castings are automatically extracted from the die face and sprayed with a release agent; conveyor belts take the castings to the trimming shop.

At first, attempts to improve the standard AZ 91 and AZ 81 alloys were made by the addition of small amounts of other elements such as calcium and silicon. It was found that with 1% of silicon and a reduced aluminium content the creep resistance was improved, although there was a tendency to hot shortness. Tests were carried out on an alloy containing 4% aluminium and 1% silicon and in 1969 the alloy termed AS 41 was introduced. It was anticipated that in the future an alloy with still higher creep resistance would be required, so a further investigation led to the alloy with only 1.9–2.5% aluminium, plus 0.7–1.2% silicon, termed as AS 21, although it was more difficult to die cast than the other alloys.

The different creep properties of the alloys are due to the fact that with increasing aluminium content the amount of the intermetallic compound $Mg_{17}Al_{12}$ increases; this has a melting point of about 460°C and is easily deformable at engine operation temperatures. In the AZ 81 alloy the intermetallic compound appears at the grain boundaries of the magnesium–aluminium solid solution.

As engine power outputs were increased, the crank cases of air cooled engines had to endure temperatures of up to 150°C, where the creep resistance of the standard alloys decreases considerably. The coefficient of thermal expansion of magnesium is about twice that of steel and so the magnesium crank cases became heavily stressed by the stud bolts, causing creep of the magnesium and consequent loosening of the bolts. Problems such as this led to the need to develop the alloys with higher creep strength referred to above.

Today highly stressed crankcases are produced in the alloy AS 21, while transmission and transaxle housings are die cast in the alloy AZ 81, similar to AZ 91.

Magnesium alloy die cast wheels

Cromodora of Turin, a major supplier of light alloy wheels to Fiat, has developed a programme for the production of magnesium wheels[6] in cooperation with the American Dow Chemical Company. Aluminium alloy wheels had been used for sporting and competition cars but there had been little experience in the use of the still lighter magnesium alloys. When these wheels were first designed in 1966, it was agreed that they should be of the wide flange type to take sporting tyres and that the inside of the wheels would have ample space to accommodate disc brakes. A magnesium alloy with 6% aluminium was selected designated AM 60A, and is still used; the alloy also contains up to 0.4% manganese, 0.3% zinc and traces of beryllium.

Preliminary pressure die castings were produced at minimum section and tested to discover the points at which cracking might occur under stress. Then the dies were opened out and fresh samples made, gradually increasing the section until it was established that the correct design had been achieved. The wheels were tested under rigorous conditions, including 'figure of eight' driving, since the greatest stresses would be involved in rapid cornering.

A research programme to improve resistance to corrosion was developed. The Dow 17 electrochemical process was used in which a coating of fluorides, phosphates and chromates was produced. Nowadays, the protective coating consists of the Dow 17 treatment followed by three layers of paint.

The castings are made on a group of 1500 and 2000 tonne Triulzi cold chamber die casting machines. The magnesium alloy is held at between 690–700°C, and the metal is protected by an atmosphere of SO_2. The runners are removed, the castings examined by X-ray, and any wheels showing evidence of porosity are rejected. Machining operations are done automatically, flash is cleared from holes, one half of the casting hub is bored, reamed and turned for tyre seating and hub contact surfaces. Then similar operations are done on the opposite half of the wheels. Next the stud holes are bored and reamed and the valve hole machined. *Figure 15.6* shows a typical magnesium die cast wheel, produced by Cromodora, adaptable for BMW and Alfa–Romeo. It weighs 5.75 kg.

Figure 15.6 Pressure die cast magnesium alloy wheel.
(Courtesy Cromodora Division of Gilardini, S.p.A.)

Other magnesium alloys

The current commercial magnesium die casting alloys already mentioned offer a combination of castability with good mechanical properties. For castings which are to be electroplated the alloy designated as AZ 61 is used; this contains 6% aluminium and 1% zinc and is similar to the composition used for automobile wheels except that it contains a higher percentage of zinc. Alloy AZ 61 is somewhat difficult to die cast, but a small increase in aluminium content to about 7% gives good castability on hot chamber machines, coupled with satisfactory plating. This alloy is identified as AZ 71.

There have been continued efforts to discover new alloys that would perform even better than those already available. For example, work in the USA was described in Die Casting Engineer by G.S. Foerster[7] and his associates. The 9% aluminium alloy AZ 91 and AM 60A with about 6% aluminium, similar to that used for the Cromodora wheels, were evaluated, followed by the Volkswagen creep-resisting alloys AS 41 and 21. Tests were then performed on the AS 21 alloy to which about 0.2% of antimony was added. This improved corrosion resistance but the alloy remained difficult to die cast.

The researchers then turned to alloys containing zinc. Although it was known that a zinc content of about 1% led to hot shortness, it was found that when the amount of zinc was increased considerably, hot cracking diminished. An alloy known as ZA 124, with 12% zinc, 4% aluminium and 0.3% manganese was developed, which proved to be suitable for die casting complex parts with thin section. Small amounts of calcium were added to ZA 124 alloy; although this element causes hot cracking in AZ 91 and similar alloys, its effect on the high zinc–aluminium alloy was beneficial. Another alloy, known as AZ 88, containing 8% of each of zinc and aluminium was developed and, in spite of a long freezing range from 577°C to 357°C, this alloy proved to be satisfactory and was recommended specially for hot chamber die casting. This research illustrates the endeavours to widen the range of magnesium die casting but up to the present time the alloys listed on page 137 are used most widely.

Pressure die casting of magnesium alloys

In both hot and cold chamber processes a fast cycling time is required to take advantage of high production rates made possible by the low heat content of the metal. For a small die casting the cavity fill time of about 0.02 seconds is appropriate. In cold chamber machines a fast acting second stage of injection is required before the onset of solidification, with a final pressure intensification of 50–70 N/mm_2^2 acting within 15–35 ms of cavity fill. Since magnesium alloys are not aggressive to ferrous components, the continued development of hot chamber production has been possible. Production speeds faster than those obtained on cold chamber machines are obtained for small and medium sized castings. Other advantages include lower metal losses and an easier approach to mechanization than with cold chamber machines.

Early developments in magnesium die casting came from Germany, where the use of the metal enabled that country to become more self sufficient. A chapter entitled 'The technique of pressure die casting'[10] was written by Dr A. Bauer for a comprehensive book on the technology of magnesium and its alloys, compiled by Adolf Beck in 1940 and soon afterwards translated into English. Bauer relates how the first magnesium die casting machine was constructed in 1925 by Elektronmetall GmbH (later to become Mahle). It was a hand-operated vertical machine and the metal was injected by compressed air; the alloy contained about 9% aluminium and was similar in composition to the modern AZ 91. Dr Bauer reported that, by 1932, 'magnesium could stand alongside aluminium die casting' and the book contains illustrations of several early magnesium alloy die castings, including one weighing 6 kg for a cash register and another weighing 4 kg for a gearbox, with 72 cored holes. By 1940 Mahle was producing magnesium die castings weighing up to 7 kg on hot chamber plunger machines. Still larger castings were made on air pressure machines, although stringent safety precautions were necessary, each machine being isolated and enclosed.

After the war, production of magnesium was reduced drastically, but technical development work continued. Dr Bauer, who had been associated with the Mahle die casting operation, moved to the USA, first to the Dow Chemical Company, then to Doehler—Jarvis. Another pioneer, F.C. Bennett[9], when at Dow, did a great deal for the development of magnesium die casting in connection with hot chamber production. In the meantime the research work of J.D. Hanawalt[19,20] led to the overcoming of many technical and metallurgical problems.

The hot chamber process became commercially viable in the mid 1960s when, among others, Oskar Frech of Germany and Idra of Italy were developing machines suitable for magnesium. In 1970 a paper by W. Frech[10] outlined the problems that had to be overcome in hot chamber production and in 1975 a further paper from the same company[11] gave details of modern hot chamber die casting and a description of their machines.

Hot chamber machines up to 700 tonnes locking force are available, but the most popular sizes are between 100 and 300 tonnes. A typical 280 tonne machine can give casting outputs of over 100 kg/hour compared to about 70 kg/hour for a similar cold chamber machine. It has been reported[12] by the German manufacturer Andreas Stihl that up to 300 shots/hour have been obtained, but the normal output ranges from 120 to 260 shots/hour. Andreas Stihl built a new plant during the 1970s for the production of chain saws using AZ 91 alloy for hot chamber pressure die castings ranging in weight from 15 g to 1 kg. Their current output is around 2000 tonnes per annum and they are engaged in building a new works in Virginia, USA for similar production.

Cold chamber machines were discarded because of the problems of automatically ladling magnesium. Equipment suitable for aluminium could not be used because the metal surface of the magnesium must be protected from oxidation while it is being moved. Hot chamber machines in which the melt is contained in the machine were more suitable for automation. A process for electric induction heating of the nozzle was developed which is less susceptible to trouble than conventional gas heating and has the advantage of enabling shorter sprues to be used, thus improving casting yield.

Numerous problems had to be overcome in applying the hot chamber process to magnesium. The stress limitations on the steel gooseneck at the casting temperatures employed and difficulties of pressure sealing the nozzle ends impose a limit of the metal pressures of not more than 2.5 tonnes/in^2 (38 N/mm^2); cold chamber machines provide up to four times that injection pressure. However, hot chamber machines work satisfactorily with pressures of 18—26 N/mm^2 and they require a locking force of only half to two-thirds that required for producing the same castings on cold chamber machines. Other magnesium hot chamber die casting problems which required individual solutions related to plunger piston rings and the material for gooseneck sleeves. The casting temperature is over 200°C higher than for zinc, so heat-resisting materials must be used. Special precipitation hardened steels have been developed such as one reported by G. Barten[15] with the following analysis: C, 0.15%; Si, 0.20%; Mn, 0.20%; Cr, 10.0%; Mo, 5.0%; Co, 10.0%; V, 0.5%.

Methods of locating and seating the die to the nozzle vary considerably, but they are all aimed at providing an efficient seal against metal pressure, easy location when mounting dies and the minimum transfer of heat between nozzle and die. If the temperature of the nozzle is too high burning of magnesium occurs between shots, with the formation of oxide deposits inside the nozzle, while if the temperature is too low, the metal freezes and prevents injection. Developments during the past decade have included hydraulic

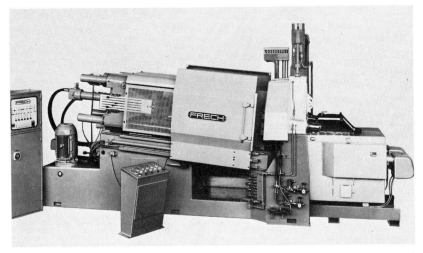

Figure 15.7 Hot chamber machine for die casting of magnesium alloys. (Courtesy Oskar Frech Werkzeugbau, W. Germany)

clamping of the nozzle to the die which allows for movements of the locking unit away from the nozzle. It is common practice to mount the locking unit at an angle, to ensure accurate line up of movements and to facilitate the back flow of metal into the gooseneck. *Figure 15.7* shows a modern hot chamber machine for magnesium.

In a paper given to the International Magnesium Association in 1979, N.C. Spare[16] described how a British die casting company designed and constructed the 'hot ends' of zinc die casting machines to equip them for the production of magnesium components. The gooseneck liners were still giving satisfactory service after 600 000 shots and the pistons lasted over 70 000 shots. Casting rates of 300 shots/hour were claimed. During 1981 a Harvill hot chamber machine was developed in the USA to provide high pressures and injection velocities comparable with those obtained in cold chamber machines.

The new technology of gate and runner design that has been such an important feature of zinc die casting has benefited magnesium, enabling shorter sprues to be used with corresponding savings in production and melting costs, and reduction of metal losses. Furthermore, as was pointed out in an address by J.A. Bolstad[15], there are other benefits of using magnesium; die life is appreciably longer than for aluminium because the alloy does not attack die steel. A die cast carburettor for the Renault 18 turbo car was described in the same address as a newly developed use of magnesium.

Cold chamber machines for magnesium

It is possible that many of the difficulties in die casting magnesium alloys arose because companies with experience in aluminium die casting did not appreciate that magnesium, with its individual properties, needs appropriate production parameters. The holding crucible in the cold chamber machine for AZ 91 alloy should be temperature controlled with a range of 650–680°C compared with 620–660°C for the hot chamber machine. The smaller evolution of heat from the solidification of magnesium offers scope for faster production than for aluminium; however the die temperature must be maintained at between 250–300°C, and it is normal practice to use internal die heating by recirculating thermal transfer fluid. The reject rate is very much dependent on die temperature, so the heating and cooling of the die is critical. Thus local hot spots need to be cooled by water lines and cores cooled by internal 'cascades' or heat pipes inside them.

In the mid 1970s, the BNF Metals Technology Centre carried out a research on the die casting of magnesium alloys, mainly concentrating on cold chamber production; their publication 'BNF Guide to Better Magnesium Diecasting'[18],gives a great deal of information on casting techniques, correction of faults and automatic metal metering. In addition the report contains advice on instrumentation, die lubrication and safety procedures.

Those who are concerned with the production of magnesium alloy die castings, by hot or cold chamber machines, agree that several requirements are necessary.

1. Uniform wall thickness.
2. Temperature control of the metal and die.
3. Continuous operation, preferably round-the-clock working.
4. Crucible cleaning several times each day.
5. Good housekeeping is even more important in die casting magnesium than with either zinc or aluminium.

Metal dispensing for magnesium alloy

Coreless induction furnaces are preferred for bulk melting where the stirring action is beneficial, although these are not suitable for holding furnaces at the die casting machine. Some manufacturers with large outputs are claiming good results with immersion tube furnaces similar to those for zinc alloy, described on page 110. Electric resistance furnaces and gas fired crucible holding furnaces are also used.

The die casting alloys do not attack mild or low-alloy steels provided that these materials do not contain nickel or cobalt. Several materials are currently used, including welded mild steel, cast low-alloy steels and heat resistant nickel-free stainless steels such as 430. A composite material, consisting of mild steel with about 2 mm of chromium–

nickel steel on the burner side has been used, giving an increased service life. Cast irons have insufficient creep resistance for use in magnesium die casting.

Several systems have been developed for dispensing magnesium alloy to cold chamber machines, ranging from pneumatic and mechanical displacement pumps to pressurized furnaces and electromagnetic propulsion of the liquid metal. Precautions need to be taken to ensure that a protective atmosphere will prevent oxidation and deter sludge formation.

The primary metal producer Norsk Hydro developed a closed unit incorporating a dosing tank which can be immersed in a normal bale-out furnace with a sealed top cover, known as the 'Normagic' dosing system. The control cabinet contains a small pressure vessel fed by argon at about $30 \, lb/in^2$ ($0.2 \, N/mm^2$) with a throttle valve on the outlet to avoid a surge of the dosing tank. On initiation of the cycle a timer-actuated solenoid valve admits argon at about $3 \, lb/in^2$ ($0.02 \, N/mm^2$) to the dosing tank. After a preset time interval, the valve closes, the displaced metal flowing along an electrically heated delivery tube and into the shot sleeve of the machine. Provision is made for a small controlled volume of argon to pass along the tube and into the shot sleeve for added protection of the metal surface before injection into the die.

Volkswagen use a direct melting and dosing furnace where charging of ingots is made through a nitrogen gas lock. A launder delivery tube attached to the furnace is pressurized with nitrogen for the time necessary to deliver each dose of metal. Electromagnetic transfer of liquid metal as discussed on page 176 has been applied to magnesium die casting and for adequate melt surface protection, the delivery launder is replaced with a sealed tube in which a protective gas can be introduced.

Metallurgical control

Magnesium is often bulk melted under a flux which protects the metal from oxidation; some processes for the treatment of molten magnesium are given on page 183. Magnesium oxide is softer than aluminium oxide, so does not cause such 'hard spot' problems as those described on page 71. However if magnesium oxide is allowed to accumulate, the fluidity of the alloy is deteriorated.

Research work by Professor J.D. Hanawalt and his co-workers at the University of Michigan[17,18], indicated that fluxing causes unnecessary extra costs, and is responsible for operator discomfort, sludge disposal problems and air pollution. The obvious way to avoid fluxing is to conduct all melting and handling operations under an inert atmosphere. This led to the beginning of a research by the Magnesium Research Center of the Columbus Laboratory of Batelle to study the use of air plus small percentages of sulphur hexafluoride (SF_6) as an alternative to flux for melting AZ 91 scrap. In the early years of magnesium pressure die casting the protective gas was generated by burning sulphur within the cover. Then sulphur dioxide was, and indeed still is, used in some foundries. However, a more efficient development in protecting magnesium alloys from oxidation has been the use of small metered concentrations of SF_6 which form an inert and impermeable layer of magnesium fluoride. Although more costly per unit weight than SO_2 the concentration requirements are so low that its use is more economic.

During the development of magnesium pressure die casting other gases apart from SO_2 and SF_6 have been used. Nitrogen is employed in the Volkswagen foundries as a pressurizing agent in metal pumps. Carbon dioxide has been used, but is only satisfactory at high concentrations of over 90 vol %; a protective atmosphere at low concentrations leads to the formation of dross consisting of MgO and finely divided carbon. Carbon tetrachloride has been mentioned as a possible material, which might also help grain refinement but, so far as the Authors can discover, it has not yet been tested adequately. Argon has been used, for example, in the operation of 'Normagic' metal metal dispensing pump and it is possible that argon will continue to be used in the future. Nevertheless, such equipment must be sealed efficiently against leakage whereas an atmosphere of air with 0.1–0.5% by volume of SF_6 does not require such complete prevention of gas leakage.

SF_6 has been known as a possible protective atmosphere for many years, but its use was not fully acclaimed until about 1970. It was first described in the chemical literature in 1902 and was listed in a US patent in 1934, along with 26 other fluorine-containing compounds suitable for inhibiting the oxidation of molten magnesium. The references to SF_6 were overlooked, and it was not till 1948 that it became available commercially and was tested as an inhibitor. Unfortunately SF_6 was discarded because it attacked ferrous equipment in the concentrations used in the experimental work in the 1950s. At high temperatures and at high concentrations SF_6 is reactive with magnesium, iron and glass, forming MgF_2, FeF_2 and SiF_4 respectively. But at much lower concentrations of SF_6 in air (in the range of a few tenths per cent) these reactions are negligible or absent and a protective impermeable film is formed on the magnesium melt surface which provides a metallic appearance even after exposure periods of more than 2 hours. In the paper by Professor Hanawalt[18] referred to above it is shown that for SF_6 concentrations in the range of a few tenths of a percent the reaction is

$$Mg + SF_6 + air \rightarrow MgO \text{ (thin protective film)}$$

but at higher concentrations of SF_6 the reaction is

$$Mg + SF_6 + air \rightarrow MgO + MgF_2 + SO_2 + MgS \text{ plus gaseous fluorine compounds.}$$

Those who have been associated with the treatment of metals will be aware of the tendency to argue that if a treatment in small doses is effective it may be even more so in large doses. This approach does seem to have caused SF_6 to be too lavishly used and wrongly condemned.

It is difficult to get exact comparisons between the cost of the alternatives, but a paper by G. Schemm[19] of Volkswagen gave information concerning four protective gases used with magnesium, showing costs in Deutschemarks, and melt losses per cent. *(Table 15.3.*

TABLE 15.3

Gas	Hourly consumption ($\mathcal{Q}$)	Hourly cost (D.M.)	Cost per 1000kg melt (D.M.)	Melt loss (%)
Nitrogen	720	0.274	1.07	2.5
Sulphur dioxide	180	0.256	1.00	1.8
Carbon dioxide	600	0.504	1.87	2.0
Sulphur hexafluoride	1	0.129	0.50	0.35

The table was based on tests concluded in 1971, but it indicated the potential economies of using SF_6 and the improved melting loss figure obtained. In the UK during 1982 a 25 kg cylinder of SO_2 cost £17 (£0.68/kg). A 9.5 kg cylinder of SF_6 cost £80 (£8.42/kg). Comparative costs in other areas vary widely and also depend on the amounts purchased but even if SF_6 cost 100 more than SO_2 it is still more economic. SF_6 is colourless, odourless and non toxic; it is about fives times heavier than air. It is liquefied under pressure and stored in gas bottles.

The later paper by J.D. Hanawalt[18] concludes with some guidelines for the successful use of SF_6 and these are summarized below.

1. Oxygen in some form must be present for the SF_6 inhibitor to function because the optimum protective film is MgO, not MgF_2.
2. If the SF_6 protection is not adequate, the natural tendency of the operator is to increase the SF_6 flow, whereas the correct solution is to keep the concentrations uniform and low.
3. Excessive surface agitation should be avoided because protection is due to the reaction of the atmosphere with the melt and is most rapid on a fresh surface.
4. Temperatures should not be any higher than necessary, since a higher temperature requires a higher percentage of SF_6 and thicker and more rapid film formation is then unavoidable.
5. For temperature in excess of about 700°C, a CO_2 atmosphere is preferred. If approximately 100% CO_2 cannot be maintained then in many situations CO_2/SF_6 may be more effective than air/SF_6.

The role of beryllium in magnesium die casting

Beryllium is only slightly soluble in magnesium but is used in small concentrations (usually 0.0015–0.005%) to minimize burning and reduce oxidation of the melt, so decreasing the possibility of inclusion entrapment in the castings. At these concentrations there is no effect on the mechanical properties of the alloy in the die cast form. The mechanism of protection arises from a modifying effect which the beryllium has on the normal oxide film, rendering this more resistant to diffusion and hence more protective against further oxidation of the underlying metal.

The beryllium is sometimes added in the form of small notches or pellets of an aluminium alloy containing either 1% or 5% beryllium, plunged in the melt and dissolved by stirring. Control must be exercised on the amount of remelt scrap used, both by analysis and observing the suppression of burning of any oxide present on the melt surface. Otherwise alloy ingots already containing beryllium are supplied. Alloys treated with beryllium are not suitable for casting by any method other than pressure die casting.

Metal losses

One of the most controversial subjects in magnesium die casting relates to the metal losses which must be anticipated. Plants which operate continuously with a high level of production and metallurgical control have metal losses of the order of 5–8%. On the other hand companies with previous experience of aluminium which decide to undertake trial orders for magnesium using cold chamber machines, with small quantities and interrupted production, find they are having metal losses of 20–25%. Any company which is beginning small scale production of magnesium die castings should allow for such heavy losses but equally they should be aware of the substantially better performance that can be expected when production improves and orders justify continuous working. In any case the metal loss in a die casting plant should never be assessed on the basis of a single melt under controlled conditions but should be measured over a period as will be discussed in Chapter 20.

Due to the tendency of magnesium alloys to oxidize when overheated there is a great deal of difference between the condition of a crucible that has held metal for rapid, continuous production and one which has been involved in slow, troublesome single shift working. Such a crucible needs a great deal of cleaning at the end of a shift and this is where heavy metal losses can occur. When production is trouble free, when double or treble shift work is possible and when metallurgical control is well supervised, the amount of dross to be removed in crucible cleaning is small and total metal losses (calculated over a period, not just measured from one crucible of metal) can be of the order of 8% – comparable with those obtained with aluminium.

The recent developments in improved runner and gate design have led to an increased casting-to-runner ratio which has improved the metal yield and, therefore, reduced metal losses. Furthermore, the utilization of hot chamber machines has made further improvements, partly through the increased weight of magnesium that can be die cast per shift and, therefore, the smaller proportion of dross to be removed in pot cleaning.

The design of magnesium die cast components and of the dies from which they are produced are not within the scope of this book but numerous publications by magnesium producers and development associations give guidelines for successful design. The International Magnesium Association, centred at Dayton, Ohio, issues newsletters and brochures on many aspects of magnesium production and use. The primary metal producer, Norsk Hydro, has several useful publications including 'Magnesium, pure and alloys', which includes information on casting, machining, finishing and safety rules. In the present book protective and plating treatments of magnesium are discussed on pages 297 and 298.

References

1.   (Editorial)
      Magnesium production hits all-time high. *Magnesium* International Magnesium Association, Dayton, Ohio (April 1981)

2.   DOWSING, R.J.
      Materials from the sea; inexhaustible source of magnesia and magnesium. *Metals and Materials*, 9–26 (January 1978)

3.   HOY-PETERSEN, N.
      Magnesium production at Porsgrunn, Norway. *Light Metal Age,* 37, (7/8), 24 (August 1979)

4.   INTERNATIONAL MAGNESIUM ASSOCIATION
      Magnesium Die Casting for Automotive Applications. International Magnesium Association, Dayton, Ohio

5.   HÖLLRIGL-ROSTA, F., JUST, E., KÖHLER, J.; MELZER, H-J.
     Magnesium in Volkswagen. An illustrated report produced by
     Volkswagen AG Zentrallaboratorium, Wolfsburg, West Germany

6.   FIORELLI, A.; PINOMONTI, C.
     Ruate in lega di magnesio per vetture spontive prodotte in serie. Reprint of a report issued by
     Societa Cromodora, Turin (1968)

7.   FOERSTER, G.S., GALLAGHER, P.C.J., HAWKE, D.L.; AQUA, E.N.
     Research in magnesium. *Die Casting Engineers,* **21,** 12 (January/February 1977)

8.   *The technology of magnesium and its alloys.* Published by F.A. Hughes and Co. Ltd., London.

9.   BENNETT, F.C.
     Die entwicklung verbesserter automatischer beschickung–vorrichtungen fur magnesium–
     druckguss. *Giesserei,* **57,** 169 (1970)

10. FRECH, W.
     Hot chamber die casting of magnesium alloys *Society of Die Casting Engineers* Congress Paper
     No. 142 (1970)

11. FRECH, W.
     Experiences and possibilities in hot chamber die casting of magnesium. *Society of Die Casting
     Engineers* Congress Paper No. G-T75 - 113 (1975)

12. MEZGER, H.
     Zeitschrift fur industrielle Fertigung. Springer-Verlag (1978)

13. BARTEN, G.
     Magnesium–Druckguss Warmkammer Erfahrungen mit einem neuen Warmarbeitsstahl fur
     Geissbehalter, Kolben und Kolbenringe. *Warmebehandlungstechnik,* p. 307 (1977)

14. SPARE, N.C.
     Magnesium die casting in Great Britain. *International Magnesium Association 36th World
     Conference on Magnesium* (May 1979)

15. BOLSTAD, J.A.
     Hot chamber magnesium die casting. *Tenth International Pressure Die Casting Conference,*
     Madrid (May 1981)

16. *BNF Guide to better Magnesium Diecasting.* BNF Metals Technology Centre, Wantage, Oxford-
     shire (1976)

17. HANAWALT, J.D.
     Practical protective atmospheres for molten magnesium. *Metals Engineering Quarterly,* **12,** (4),
     6–10 (November 1972)

18. HANAWALT, J.D.
     SF_6 – protective atmosphere for molten magnesium. *Society of Die Casting Engineers* Congress
     Paper No. G-T75 - 111 (1975)

19. SCHEMM, G.
     'Sulphur hexafluoride as a protection against oxidation of magnesium melts' (in German and
     English) *Giesserei* p. 559-656 (23rd September 1971)

Chapter 16

Brass die casting

The world's production of copper is 8–10 million tonnes per annum, third only to iron and aluminium. In spite of competition from aluminium about half of all the copper produced goes to the electrical industry and it is expected that over 60% will be required for those purposes by the year 2000.

Copper ores are distributed widely. The USA contains the largest reserves, equivalent to over 80 million tonnes of the metal; Canada has 30 million tonnes, so for the present North America is self sufficient. Soviet Russia contains large reserves in Kazakhstan, Dagestan and in the Urals. Chile, producing over 10% of the world's copper, was one of the founder members of CIPEC (Conseil Intergouvernmental des Pays Exporteurs de Cuivre), an organization which was formed to protect the interests of some copper producers[1]. Other members include Peru, Zambia, and Zaire. Australia, Papua and New Guinea are associated with CIPEC. These countries together account for nearly half the world's copper requirements; the USA, Canada, South Africa and West Europe produce about a third; the USSR and its associates the remainder, about 20%.

Up to the early twentieth century copper was extracted from ore bodies sufficiently rich to be hand picked but today any ore with over 2% metal is regarded as high grade, while most ores contain less than 1%. The principal copper ores contain sulphides, either of copper alone or of copper and iron such as chalcopyrite, which yields about half the world's copper. Other ores contain carbonates and oxides. Whether the ores are obtained from open-cast mines or underground, copper-containing mineral must be separated from the stony mass with which it is associated. Giant crushers to bring the rock to 15cm lumps are sometimes installed down the mine; the ore is crushed to a smaller size at the surface, then ground to a powder. The copper-rich mineral is separated by flotation.

Copper requires only about one-eighth the amount of energy per tonne of metal compared with aluminium, but the need to work ores of ever-decreasing richness is leading to an increase of energy consumption. Although there will be a shortage of metal by 2000 AD, three factors must be considered before becoming unduly pessimistic. Since copper is resistant to corrosion and since most of its uses are for products which can be reclaimed after many years service, secondary copper provides a large source of the world's requirements, amounting to about 35%. A great deal of work is being done to develop new processes[2], for example the Outokumpa flash-melting process, invented in Finland, to reduce the energy required per tonne[3]. Much attention has been given to developing electrochemical methods of exploiting weak ores and residues. One potential source of the metal comes from tailings of previous years which were then thought useless but now provide a valuable supply. The solvent extraction process depends on a series of reactions in which organic chemicals dissolve the metallic compound, which is separated; then the process is reversed so that the organic compound is reclaimed and the liquid containing the copper goes on for electrolytic separation. Various proprietary solvents are available and extraction plants are operating in the USA and Zambia with a potential output of over 200000 tonnes per annum.

During the past decade much interest has been shown in the metallic nodules which lie at the seabed in several areas, including some to the west of California. These contain manganese, nickel and copper. Unfortunately the deposits which have been identified are all at depths of 3000 m or more, and the problems of dredging from such a distance are formidable. Also the political consideration of who owns the sea bed will have to be determined before fruitful international efforts can be made to win these rich sources of metal, but the nearer the time approaches when mines on land become exhausted, the more enterprise will be devoted to deep sea mining.

Because ore deposits will become depleted within less than 20 years and the cost of energy will increase, world-wide efforts have been made to win copper in new ways and to reduce the energy cost as much as possible. The conservation of materials will be just as important as the conservation of energy. The techniques of 'thin wall die casting', leading to notable decreases in the weight of runner systems will save metal and help die casting to compete strongly with other methods of shaping brass. Furthermore, the fact that a die casting provides a close approach to the finished product means that less machining and swarf will be involved than in competing processes.

Metallurgy of die casting brasses[4]

The suitability of an alloy for die casting is dependent on its melting point and fluidity. For any range of alloys maximum fluidity is achieved from eutectic compositions or alloys which solidify over a short freezing range and the least 'die castability' is obtained from alloys with a long freezing range. For example, in the copper—tin alloys the composition with 13.5% tin has a freezing range of 170°C; such an alloy can be cast by sand or permanent mould methods where comparatively slow rates of cooling allow progressive solidification, reducing the risk of trapping pockets of solidifying metal, but it would not be suitable for pressure die casting.

High melting point is another factor which makes die casting difficult; consequently pure copper and alloys with over 90% copper content have been die cast only experimentally. Brasses of copper with over 35% zinc are the most widely used for die casting, partly because their melting points are comparatively low, their freezing range not too wide and their strength at temperature below the freezing point is sufficiently great to withstand the stresses encountered during solidification.

A section of the equilibrium diagram of the binary copper—zinc alloys in *Figure 16.1* shows that an alloy with 70% copper and 30% zinc solidifies as an α solid solution and with a freezing range of about 30°C. As the copper content increases, the resistance to corrosion improves, but the melting point increases. However the copper—zinc alloys which solidify with the α structure are hot short. Consequently, although

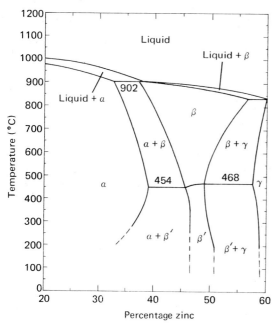

Figure 16.1 Portion of the equilibrium diagram of the copper–zinc alloys

suitable for sand and shell mouldings, they are prone to shrinkage cracking if cast in metal moulds. The alloys with about 40% zinc solidify with a short freezing range and form an α–β structure which is ductile at temperatures near solidification; they also have a lower melting point of about 900°C. Above 45% zinc the alloy contains more of the β constituent, which is brittle. Consequently a composition with about 40% zinc is most suitable for die casting. Small quantities of other elements are added.

Alloying elements

Manufacturers and users of brass die castings must compromise between the ideals of strength and maximum resistance to corrosion, against the best suitability for the die casting process[5]. Alloys with high copper content possessing an α structure are more resistant to corrosion than an α–β alloy, which is easier to die cast. Therefore most brass die castings are produced in compositions based on the zinc content of 35–40%, although often variations of the standard alloy are used; for example the American alloy CA 879 contains 65% copper, 1% silicon and the balance is zinc.

Lead is practically insoluble in copper alloys but, when dispersed through the structure, it improves the machinability of the castings, although its concentration must be limited to avoid loss of strength at elevated temperatures. In the British die casting brass DCB 3, lead is specified as from 0.5–2.5%. The Swedish A-Metal combines about 2.5% lead, 1% tin, 0.3% aluminium and 0.5–0.9% silicon. It is claimed that the silicon content improves the alloy's fluidity, which is important in the production of thin-walled castings; it also increases wear resistance of the alloy. If silicon were the only additional alloying element, machinability would be prejudiced, but the lead content overcomes that disadvantage.

Both silicon and aluminium in small concentrations reduce oxidation of zinc from the melt and minimize deposit build up on die cavity faces. Aluminium in excess of specification may lead to higher solidification shrinkage and encourage the formation of aluminium oxide inclusions, so it is usually specified as below 0.8%. Sometimes up to 1.7% is included[6] to obtain improved resistance to wear, as for example the French alloy UZ 27 M3 A$_2^2$ is used for automobile gear selector forks.

Antimony and arsenic in small amounts improve corrosion resistance but of the two elements arsenic is preferred, because even a small amount of antimony can lead to solidification cracking. Tin assists in isolating any β phase which is present and, in a similar way to nickel, it can improve mechanical properties.

Manganese adds toughness and increases the strength of copper alloys and its presence increases the amount of iron which can be taken into solution. Alloys named high tensile brasses or manganese bronzes are typical. For example the alloy BS 1400 HTBI contains 55% copper, 3.0% manganese, 0.7–2.0% iron, 1.0% nickel, small amounts of tin and lead and the balance zinc. In Britain this alloy is used mainly as permanent mould castings. In the USA one die cast manganese bronze, CA 863, contains 64% copper, 3% manganese, 3% iron and 4% aluminium.

Brass die casting alloys

The literature on brass die casting contains many references to alloys that have been developed in the efforts to combine strength, resistance to corrosion and die castability. They may be divided into the following groups.

1. General purpose alloys containing from 57–65% copper, aluminium usually under 0.8%, lead up to about 2.5% tin up to 1.0% and small allowable amounts of nickel, and iron. In some variations of the standard alloy, up to 1% silicon is included. The two American specifications CA 879 and CA858 are typical of this group and in Britain DCB 3 is the standard alloy.
2. Copper from 64–65%, lead about 2% plus small amounts of tin, manganese and silicon, with the balance zinc. The Swedish 'A-Metal' mentioned above comes into this group.
3. Brasses with a high copper content, about 82%; silicon about 4% and the balance zinc. Such alloys have high strength.
4. Copper about 60%, manganese 13–20%, aluminium about 1.25%, lead 1.5% and zinc balance. The proprietary alloys Tombasil and Bronwite come within this group; they are silvery in colour. Tombasil contains about 5% nickel in addition to the other elements.

Table 16.1 shows the composition of the British alloy BS 1400, DCB 3, the brass widely used for die cast water fittings. Typical compositions of the American alloys CA 879 and 858 are also shown and, although in some respects their compositions are similar to DCB 3, it will be noted that the British designation allows a greater percentage of lead.

TABLE 16.1

	BS 1400 DCB 3 (%)	CA 879 (ZS 331A) (%)	CA 858 (Z 30 A) (%)
Copper	58-63	65	58
Tin	1.0	0.25	1.0
Lead	0.50-2.5	0.25	1.0
Iron	0.5	0.15	0.50
Aluminium	0.2-0.8	0.15	0.25
Manganese	0.5	0.15	0.25
Silicon	0.5	1.0	0.2

Die casting production

Copper alloys have a combination of high strength, resistance to corrosion and an attractive range of natural colours. Therefore they offer good possibilities to the die casting industry, although the alloy cost per unit volume and the need for heat-resisting die materials and specialized production techniques mean that brass die castings compete in different markets from those covered by zinc, aluminium and magnesium.

Although copper alloys were die cast on cold chamber machines in the late 1920s, progress was slower than with other metals, due to the higher casting temperatures and relatively short die lives. As a general rule, companies with a large output of zinc or aluminium have found difficulty in 'coming to terms' with the special requirements of a small output of brass.

The main developments have come in two directions. During the late 1950s brass die casting was established in several countries as 'cottage industries'. For example, the village of Lumezzane near Brescia in Italy and Gnosjö, to the south of Jönköping in Sweden have numerous small workshops specializing in making brass die castings. In Israel one small company[8] achieved remarkable results in extended die lives and high productivity by improving die making techniques and developing automation. Such communities operate only a few machines, producing mostly water fittings and it is likely that their success came because, not being involved with lower melting point metals, they had few preconceived ideas. Their prosperity depended on success or failure of the one branch of the industry which they had selected.

Elsewhere there are several specialist companies, and some plants which have developed as branches of very large concerns, but in every case the chances of success in brass die casting appear to be better if the plant concentrates on that one field of production. These large and small die casting companies manufacture water fittings and their high standards of production and aggressive marketing enables them to be competitive against sand castings and hot brass stampings, though mechanized permanent mould casting is still a powerful rival. Other brass die castings include door and furniture fittings, typewriter components, lock parts and electrical fittings, some of which are illustrated in *Figures 16.2* and *16.3*. *Figure 16.4* illustrates the saving that can be made by coring holes; the circular casting is an end plate for a heat exchanger. This brass die casting has been cut to show the thin section that can be cast. *Figure 16.5* shows a group of automobile gear selector forks. A view of the brass die casting plant of FAVI, in France,

Figure 16.2 Brass die castings including water fittings, electrical components and a welding clamp. (Courtesy FAVI.s.a)

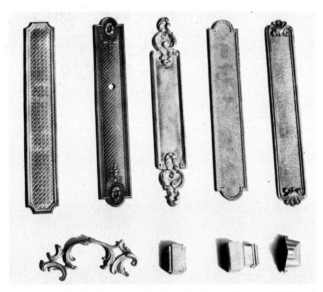

Figure 16.3 Brass die castings for door furniture. (Courtesy FAVI.s.a.)

one of the world's largest specialist producers of brass die castings, is shown in *Figure 16.6.* The machines in the foreground are 150 tonne capacity and 225 tonne in the rear. The FOMET double chamber 120 kW furnaces are used for both melting and holding.

Most brass die casting plants are developing mechanized production[9]. Following the automatic ladling of metal to the machine, castings are removed from the die cavity by extractory devices or a robot. Runner and overflow areas are automatically cropped and the castings trimmed. Mechanized die casting, with its regular die cycling and even die temperature, leads to a uniform casting quality, less oxide build up on the dies and long die lives. Avoiding turbulence during ladling is important, to reduce the rate of oxidation on the surface of the melt. Crazed and worn die surfaces can make casting withdrawal difficult, because, unlike a skilled operator, extractors and robots cannot sense the need for manoeuvring the casting away from the die face.

Figure 16.4 A heat exchanger end plate with many cored holes, water fittings, showing thin section. (Courtesy J.W. Singer & Sons Ltd.)

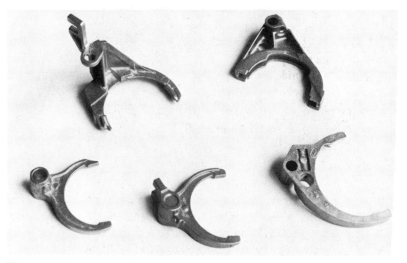

Figure 16.5 Brass die cast automobile gear selector forks. (Courtesy FAVI.s.a.)

Figure 16.6 A view of the FAVI brass die casting plant at Hallencourt

Dezincification

The α–β brasses are susceptible to a form of degradation called dezincification[10] when in contact with aggressive fluids, or even water in certain areas. This effect is shown as a slow dissolution of both copper and zinc from the brass which leaves a porous copper rich constituent that will fracture easily when stressed. It is generally believed that following dezincification, copper is redeposited at the site of corrosion while insoluble zinc compounds, for example zinc carbonate, may also form. In duplex alloys used in die casting, the β phase is susceptible to this form of corrosion, depending on many factors including the water's chloride content, acidity level, concentration of dissolved gases and temperature. Thus the problem of dezincification affects the choice of alloy composition for castings used as water fittings and has led to the development of alloys which do not deteriorate.

Metallurgical research to develop brasses that resist dezincification has concentrated on manipulating the microstructure in such a way that the β phase is distributed non-continuously within the α phase. Any corrosion which may occur would then be confined to the surface and would not develop beyond that area, since protection is offered by the surrounding α matrix. A small amount of arsenic acts as an inhibitor against selective corrosion of the α phase. Antimony serves in the same way but its content is maintained at a minimum level to prevent solidification cracking of thin-walled castings. A number of alloys have been introduced to resist dezincification by isolation of the β phase. A-Metal, developed by A.H. Andersson[11] contains a reduced amount of β phase due to an increase in copper content from 60% to about 65%, while still retaining an $\alpha-\beta$ phast to assist castability. In France[6], the alloy ND2, with 65% Cu, 1% Si, 1% Pb and traces antimony and arsenic, was developed to resist dezincification.

Accelerated testing methods have been developed to measure resistance to dezincification. Specimens are exposed for a fixed time and temperature to chemical solutions and then examined visually and metallographically. For works' investigations a simple test can be made during casting production with a few drops of sulphuric acid: discolouration indicates a tendency to dezincification.

Zinc oxide build up

When brass is melted, zinc vapour is formed which oxidizes and produces a hard tenacious scale in combination with die lubricant. Zinc oxide is an insulator which reduces the thermal conductivity of the die surface; the oxide builds up quite rapidly and affects the accuracy and appearance of the die casting. Furthermore, any surface cracks act as anchoring points which retain oxide deposits. Usually dies are taken from the machines and scale is removed by shot or sand blasting, but frequent treatment may affect die dimensions or cause corners to be rounded.

An alternative method of die cleaning requires the use of chromic acid solutions which are effective on die steels but can damage TZM cores. The die is partly cooled by leaving water circulating through it for a quarter of an hour at the end of the shift. Then the die is taken from the machine and sprayed with water until it is completely cool. The solution is poured into the die cavity where it remains overnight. Before the next day's production the die is cleaned with water, and the released oxides wiped away. If the configuration of the die is not suitable for this treatment, the whole die section can be immersed in the cleaning solution.

Another effective treatment involves adding a small amount of magnesium to the molten brass in the ladle several times each hour. The few castings produced with this addition are discarded but the magnesium has assisted in removing oxide from the die surface. Care must be taken with old dies because the magnesium will remove deposits of material from deep surface cracks, leading to difficult release of castings produced after the treatment.

Metallurgical control

During the melting of brass strict control must be exercised if loss of zinc by volatilization is to be minimized. Reverberatory furnaces where the flame of combustion is above the molten metal will lead to heat concentrations on the melt surface, increasing the loss of zinc. Use of lightweight finely divided scrap such as turnings and sweepings will accelerate metal losses since the exposed surface tends to oxidize as it melts in the alloy bath. Any zinc loss in melting will require the addition of pure zinc to the charge to ensure that metal of the correct specification is held at the die casting station. Temperature control is important to ensure that metal losses are minimized, to prolong die life and to encourage high productivity. Only a limited amount of fluxing and degassing treatments are necessary, particularly when electric melting is used.

Electric induction furnaces[12] for melting and holding brass alloys result in a smaller loss of zinc than in oil or gas fired furnaces. In addition to the obvious advantages for metallurgical and processing aspects, modern electric furnaces are economical in use and clean in operation. Where frequent alloy changes are needed the twin chamber type of induction furnace[13] is often preferred. One chamber is used for melting, the other for holding with each connected below the surface of the metal in such a way that the chilling effect of charged material is confined to the melting chamber.

Die and core materials

Much effort has been devoted to finding ways of prolonging die life. The importance of correct die and core material for aluminium alloy die castings in particular will be discussed in Chapters 25 and 26, but these subjects have even more significance when applied to brass die casting. Although thermal fatigue, leading to craze cracking is the main cause of reduced die life, mechanical erosion by the molten alloy, the metal's adhesion to the die surface and oxide build up all contribute to accelerated die wear. Although the 5% chromium die steel H 13 has been used, other steels such as H 10A, 19 or 21 offer improved endurance for brass die casting. Special heat resisting alloys are used widely for die inserts.

Omen Metal Products[8] in Israel, mentioned on page 153 attributes their progress to the use of improved die materials. An article in Precision Metal[14] also reports the unusually long die lives which they achieve, and shows illustrations of some aspects of Omen's production. At first they used a chromium–molybdenum–cobalt–vanadium steel similar to H 10A, heat-treated to a hardness of 42–44 HRC. This steel is capable of retaining its hardness at high temperatures; for example at 620°C, HRC 48; at 650°C, HRC 46 and at 680°C, HRC 40. Using this steel the Israel company[14] claims to have produced more than 300 000 shots from the same 12 impression die shown in *Figure 16.7*. The die operates at a temperature of about 400°C.

Figure 16.7 Twelve–impression die for brass die castings. (Courtesy Omen Metal Products.)

There is a growing use of refractory metals based on molybdenum and tungsten, discussed in Chapter 25. Refractory metals offering a high rate of heat removal enable the speed of casting production to be increased which leads to better temperature distribution on all parts of the die.

The selection of suitable die material is the first step to long die life, but the conditions of the casting operation must be monitored and controlled to provide the minimum temperature range on the die surface between metal injection and the moment before the next casting. It is also important that the dies are preheated close to their working temperature before starting to cast, to avoid the sudden heat shock when molten metal is injected at about 950°C. Successful brass die casting operations require the postponement of heat checking by preventing the die surface temperature from suffering wide temperature variations.

Some die casting companies are stress relieving their dies between production runs and find that die lives can be prolonged considerably by doing so. Product design is of equal importance, since it leads to casting conditions required for minimum temperature gradients. Uniform section thickness, preferably around 1.5mm gives the best conditions and the paper[8] referred to above states that castings which do not fulfil these conditions give early die failure. Gates and runners are of minimum size — a technique which is now accepted throughout the die casting industry. In designing the die, uniform heat distribution is necessary and water cooling channels are located as far as possible from the working sections so that the die temperature is not allowed to fall below 450°C. Casting rates averaging 170 shots/hour are reported in the same paper, but as many as 250 shots/hour can be achieved.

Although less has been written about brass die casting than of zinc and aluminium, two publications provide comprehensive information on product and die design, die materials, casting techniques and lubrication. 'Better Brass Diecastings'[15] issued by the BNF Metals Technology Centre in 1972, was one of the first reports to link a scientific analysis of casting conditions with the quality of the die cast product. 'Copper Alloy Die Pressure Casting'[4] was produced by the International Copper Research Association in association with the Society of Die Casting Engineers. An informative paper on the future for brass die casting was given to the First South Pacific Diecasting Congress held in Melbourne in 1980[16]. All are well illustrated with pictures and diagrams and contain references to technical papers on brass die casting.

References

1.   DOWSING, R.J.
     Copper: the enduring metal adapts to meet changing markets. *Metals and Materials*, **10**, 19–25 (June 1976)

2.   DOWDING, M.F.
     The world of Metals. Presidential address to the Metals Society, *Metals and Materials*, **12**, 27–36 (July 1978)

3.   BISWAS, A.K.; DAVENPORT, W.G.
     Extraction metallurgy of copper. Pergamon Press, Oxford (1976)

4.   MACHONIS, A.A. (Editor) HERMAN, E.A.; WALLACE, J.F.
     Copper alloy pressure die casting. International Copper Research Association, Inc., New York (1975)

5. STAMFORD, M.S.
    Copper alloys for diecasting. *The Diecasting Society (UK) Annual Conference.* Zinc Development Association, London (March 1979)

6. TOULEMONDE, D.
    Private communication from FAVI, s.a., Hallencourt, France.

7. MACHONIS, A.A.
    Copper alloys and the die casting industry. *Die Casting Engineer,* **24,** (6), 38 (November/ December 1980)

8. SCHWARTZ, M.
    On the subject of brass die casting. Omen Metal Products, Israel

9. HARRISON, R.
    Brass pressure diecasting. *British Foundryman,* **66,** 23 (January 1975)

10. (Editorial)
    Better alpha–beta brasses for pressure diecasting. *Diecasting and Metal Moulding,* **8,** (11) (September/October 1978)

11. SVENSSON, A.
    Die casting of corrosion–resistant copper alloys. *Society of Die Casting Engineers.* Congress Paper No. 9472 (1972)

12. SUNDEEN, R.W.
    Induction melting for brass and bronze. *Foundry,* **97,** (11), 60 (November 1969)

13. FRENCH ELECTRICITY AUTHORITY
    La fusion électrique du laiton au four a induction a canal bi–bassins a la FAVI. *L'Electricité dans l'industrie* (April/May 1976)

14. (Editorial)
    Die casting brass? 340 000 shots per set of tools. *Precision Metal,* **28,** 72 (November 1970)

15. HILL, T.B.; BATES, A.P.
    BNF Guide to better brass diecasting. BNF Metals Technology Centre, Wantage, Oxfordshire (1972)

16. HERBST, N.
    The Future of Diecasting. First South Pacific Congress, Melbourne, Australia, 1980. Society of Diecasting Engineers, London

Chapter 17

Ferrous die casting

Those who have been immersed in an industry such as die casting often make impressive developments; the gradual progress from the early machine that needed the efforts of four men to make a casting to the automatic and computer-controlled die casting machine of today illustrates what has been achieved step by step within the industry. On the other hand, when a completely new approach is required it is sometimes easier for those outside the existing industry to invent new ways of tackling new problems. Most die casters who are involved with alloys of low and medium melting point would know that die lives for zinc run into hundreds of thousands and aluminium tens of thousands. They might then extrapolate the temperature-die life graph and come to the conclusion that if die casting in steel were attempted, the die life would be a handful of castings of doubtful quality. However, by taking a 'new look' at the problem, the pressure die casting of ferrous metals has been accomplished. Furthermore, pure copper and some of its alloys including aluminium bronze have been die cast, although that was previously thought to be impracticable.

The casting of ferrous metals in permanent moulds is by no means unusual. Most pressure pipes for gas and water are centrifugally cast in metal moulds by the De Lavaud process, using a powder dressing on the mould surface. Grey cast iron and, to a lesser extent, spheroidal graphitic iron, is permanent mould cast by the Eaton process, described in an article in *Modern Casting*[1]. One foundry in Britain produces over 300 tonnes per week for such parts as automobile brake cylinders. There are three large manufacturers in Europe and several in the USA.

The early efforts in ferrous pressure die casting go back to the mid 1960s, when experimental production was carried out in many parts of the world, including the USA, Japan and Russia[2]. It was realized that a new approach would be needed if the project was to be fulfilled, but early experiments were based on the routine of die casting of medium melting point alloys, leading to many frustrating problems.

In one series of experiments in Britain, four men were involved in operating a machine, reminiscent of Doehler's experiments of the first decade of this century, described on page 3. The machine was an old 'Cleveland' and the first dies were made of mild steel. One man ladled molten steel into the shot sleeve, the second received the ladle that had just been used, cleaned the skull from it and recoated it with a refractory wash. The third man made sure that enough ladles were dried and properly coated; then he passed a suitable ladle to the first operator. Molten steel was injected into the die and a fourth operator removed the finished casting. The sequence of operations was fast and at one stage it was reported that a rate equivalent to 200 castings/hour were being made, but unfortunately the production run lasted only 12 minutes.

Having seen that ferrous die casting might be possible, the various technical problems were analysed and these fell into three groups, each of which demanded a new solution[3]. A die material had to be found to withstand the thermal shock of steel injected under pressure at a temperature of around 1600°C, yet that material must not cost so much that ferrous die castings would never become competitive. It soon became apparent that conventional methods of ladling molten metal into a shot sleeve would not be suitable, so a new way had to be found for injecting the steel into the die without the excessive amount of labour and frequent breakdowns that had been experienced in the first experiments. Thirdly a reliable system had to be developed for melting and dispensing accurate quantities of molten metal at high temperature.

There were other associated problems, such as the design of components and dies, the coring of holes and recesses, the ejection of castings from the dies and the possibilities of marketing the ferrous die castings in sufficient volume to make the programme of costly development worth while. Until 1976 the work in Britain was performed at a Research and Development centre of GKN. Then a pilot plant called 'GKN Ferro–Di' was established on a commercial basis and operated till 1981 when the rights of the manufacturing process were sold to Sealed Motor Construction Company Limited of Bridgwater, Somerset. Although at first the process has been confined to their own in-house manufacturing, the company intend to develop it further, with an emphasis on the pressure die casting of electrical grade copper.

Development of the die material

Progress in ferrous die casting would have been impossible without a suitable heat resisting die material. A wide range of substances was investigated, including nickel, cobalt and tungsten-based alloys, ceramics, cermets and refractory metal coating sprayed on to steel die cavities. The final solution was made possible owing to the NASA space programme, which had been using molybdenum, a metal with the very high melting point of 2625°C, exceeded only by four other metals, tungsten, rhenium, tantalum and osmium. It is mechanically strong, much ligher than tungsten and it has a thermal conductivity about four times that of steel. The metal was readily available from the USA and Canada, as well as from Chile and Peru.

Molybdenum is sufficiently refractory to resist erosion from liquid ferrous alloys and is resistant to thermal shock, because of its high thermal diffusivity and low coefficient of expansion. Both strength and toughness are improved by cold working, but at moderately elevated temperatures, recrystallization occurs in the pure metal, and the material becomes soft and relatively brittle. The breakthrough came with the development of a molybdenum alloy with additions of 0.5% titanium, 0.08% zirconium and 0.015% carbon, known commercially as TZM. The alloy has a much higher recrystallization temperature than pure molybdenum and therefore the benefits of work-hardening can be retained at the temperatures encountered in ferrous die casting. (A computer simulation indicated that temperatures in the range of 1000–1200°C are reached momentarily by the immediate subsurface layers of the die cavity when molten steel is injected). The alloy additions present in TZM provide a degree of solid solution hardening, and dispersion-hardening results from the precipitation of complex Ti–Zr–Mo carbide particles. This finely dispersed carbide phase also helps to improve the toughness of TZM over that of unalloyed molybdenum, which contains coarser Mo_2C particles.

On economic grounds, a complete die block cannot be made of such an expensive material as TZM so it must be used as an insert, in the same way that alloy steel inserts are embodied in medium carbon die blocks for producing aluminium alloy die castings. This presented additional problems, because the coefficient of expansion of TZM is only $4.9 \times 10^{-6}/°C$ compared with $14.7 \times 10^{-6}/°C$ for medium carbon steel. Therefore special techniques had to be developed to locate the TZM inserts in such a way that the differential expansion did not produce component mismatch as the die temperature varied.

TZM is not without its drawbacks. Like some other materials with a body centred cubic lattice structure. TZM exhibits a ductile-to-brittle transition temperature which can vary widely according to the degree of working, the crystal texture and the level of interstitial impurities. To ensure that the material is always used in its ductile condition, and to minimize the effects of thermal cycling, it would be desirable to operate the dies at elevated temperatures, possibly even as high as 600–800°C. Unfortunately, however, molybdenum begins to oxidize rapidly at temperatures above 450°C – indeed because molybdenum trioxide is volatile, the oxidation is catastrophic. Therefore it is not possible to operate the dies at base temperatures much in excess of 350–400°C.

In spite of these problems, molybdenum, as represented by the TZM alloy, possesses the essentials of strength at high temperature and high thermal conductivity, which ensures that the heat of the injected metal dissipates so rapidly that the working temperature of the die insert is kept low. It is fairly soft, having a hardness of about 240 HV, so it can be machined with ease, and it is available at a price which enables the producer of ferrous die castings to anticipate that the products will be competitive.

Dies for simple steel components have lasted over 5000 shots and then have been resunk, thus saving in die replacement cost. Several factors contributed to the achievement of maximum die life: the dies were cleaned by air blasting and coated between shots with a parting agent, usually acetylene soot. The dies were not generally water cooled but an efficient system of die heating prior to a production run was necessary.

Cores and ejectors

Although the configuration of the die enables heat to escape rapidly, a core is a different matter; it is surrounded by the injected metal for a period of a few seconds and cannot dissipate the heat, so is liable to rapid failure. At the present stage of development only short cored holes more than 15 mm diameter and under 20 mm in length could be attempted with metal cores.

Developments of the Ferro-Di process included the use of expendable cores. At first ceramic materials were moulded by outside specialists, and these are still recommended for components which require a high quality surface finish and close tolerances. Such cores have the advantage over those in metal that they do not require taper and they may

Figure 17.1 Ferrous die castings with associated cores. The ball core
is of resin sand; the others are of ceramic materials

include re-entrant features, in which case the core is removed from the casting by leaching
in a solution of sodium hydroxide.

A further development has been the use of inexpensive resin-bonded sand cores
which can be made in the die casting plant and subsequently removed from the finished
casting by vibration. The cores were made with a projection so that they could be located
firmly in the die prior to the injection of the molten metal. *Figure 17.1* shows die castings
made with resin-bonded and ceramic cores. The design of a reliable system for ejecting
the casting from the die involved patented developments of the material used and the
construction of the ejectors.

The injection system

The problem of the injection system caused many early workers in the field to
abandon their development of ferrous die casting. As was described on page 162 the
early efforts to cast steel using ladles involved so much labour and effort that it became
apparent that a new system must be devised. The shot sleeve must accommodate metal at
a temperature of about $1600°C$ and withstand severe thermal gradients which cause
distortion. The plunger tip must repeatedly inject molten metal without wearing or
seizing. For the Ferro-Di process a segmented shot sleeve was developed to minimize the
tendency to distortion. Several materials were used, each selected to accommodate the
conditions specific to that part of the system.

Molten metal supply

When it was realized that ladling would not be practicable, other systems were
investigated until Measured Slug Melting (MSM) was finally developed. First a one-off
crucible with a hole in the bottom is moulded from resin-bonded sand. A cylindrical slug
of feedstock of predetermined weight and suitable diameter is placed inside the crucible
which rests within an induction coil operating at 10 000 cycles. The coil is designed so

that the metal is melted progressively downwards while the bottom of the slug acts as a seal to delay the flow of the molten steel through the bottom hole until all of it is melted. When this takes place the metal runs into the shot sleeve of the die casting machine and is immediately inserted into the die cavity. The slugs can be cut from rolled round bar, although for some stainless steels, for example grade 316, ready made slugs of the required length and diameter tend to be more economical. In a recent development, feedstock costs were lowered by shortening the sawn or precast slug to allow topping up with preweighed alloying elements or scrap, for example that which has been obtained during subsequent trimming operations.

The melting time varies between 25 and 75 seconds depending on slug size. A safety feature prevents complete melting of the metal if for example an unexpected delay occurs during the production cycle. A signal is given about 15 seconds before the completion of melting and if the operator realizes that he must not allow the melting to be completed, which would cause metal to spill, he can halt the process.

A typical production cycle, after the removal of the previous casting from the opened die involves coating the die surfaces with soot, using an acetylene cylinder without oxygen adjacent to the machine and lighting the flame from a burner. Next any expendable cores are placed in the die cavity, locating them with the projections that have been described above. Then the die is closed. Prior to die coating the operator has placed a resin sand crucible, preloaded with a metal slug in position, inside the induction coil. Then he switches on the current, so that melting is proceeding while he is dressing the die. There is a mirror above the metal container so that he can observe when the metal melting sequence, having reached the bottom of the slug, allows the molten steel to run through the hole into the injection cylinder. The die is then opened, the ejectors push the casting from the moveable die face, the casting is removed, the bonded sand container replaced with a new one and the cycle is repeated for the next casting. Standard cold chamber machines of 280 tonnes locking force were used, with the necessary alterations to accommodate the MSM melting system, and with the special shot sleeve and injection piston. With this size of machine stainless steel has been cast commercially with a maximum of 2 kg, but for experimental purposes castings up to 8 kg were produced. The process is suitable either for single castings within the weight range mentioned above or, if small castings are needed, they can be produced as a spray with up to 20 castings, in a similar manner to the die designs employed in the die casting of small non-ferrous components. The longest production runs at Ferro-Di were of small castings.

The materials that are die cast are harder than the non-ferrous alloys and in the early years of development trimming problems were encountered. Careful tool design was needed to limit the amount of power required to trim ferrous die castings. The fettling of thin flash and the witness left after removing the runner was done with normal grinding wheels and linishers. The oxide layer on the casting was usually removed by barrelling.

As with all die casting processes, cooperation between the user and producer is necessary to obtain the best results. The spray of die castings must fit within an area of 600 x 230 mm within a depth of 75 mm. The minimum section and radius must be 1 mm and generally a draft angle of 5 degrees is requested. The limits of accuracy are specified as $\pm$ 0.1 mm per 25 mm. As a rough guide, a minimum of 5000 shots begins to be an economic quantity but for experimental purposes, requiring only a few prototypes, a simple mild steel die can be made and used.

Prospects for the future

It will be gathered from the above description that the ferrous die casting process has undergone setbacks in the effort to achieve profitability, although several of the technical problems have been overcome. In the future the sales departments of ferrous die casting companies will be at a somewhat similar stage to the early developments of non-ferrous die casting marketing. They have the advantage that the vast field of ferrous components can be investigated; obviously the parts that will be made in bulk in the future are those which require costly machining to arrive at the finished shape — where the die casting can reduce much of that cost. Secondly the parts must be technically suitable from the point of view of weight and design; although, as explained above, cored holes can be made with metal, ceramic or resin-bonded sand, there are still many design limitations[4].

Just as the early die casting industry would produce batches of castings in many different alloys in their endeavours to widen the market but eventually settled down to quite a small number of alloys, so it is likely that eventually the ferrous die casting industry will consolidate similarly. So far they have cast a range of stainless steels, mostly in the austenitic 302/304 and 316 grades, and in the martensitic group they cast the 410 and 431 grades. Some components have been cast experimentally in Nimonics and the cobalt based Stellites. Pure copper and dilute copper alloys have been die cast and will offer an interesting field of application in the electrical industry. Aluminium bronze has so far been considered to be an alloy eminently suited to permanent mould casting but quite unsuitable for pressure die casting (partly owing to its high melting point and partly due to its propensity to become entangled with oxide when pressure die cast). The special features of the Ferro-Di process overcome the problem of high melting point and the characteristic type of injection appears to minimize the inclusion of oxide. *Figure 17.2* shows some die castings in steel, copper, aluminium bronze and copper—chromium alloy.

Figure 17.2 Die castings in chromium — copper, aluminium bronze, copper and stainless steel (Courtesy GKN Ltd.)

When more technical and commercial problems have been overcome, the way should be open for marketing studies to indicate the area where the special features of ferrous die casting, and its associated casting of other high melting point alloys, can enable this comparatively new industry to expand into wider commercial, and eventually profitable, production. When a company is operating pilot plants, requiring a combination of technical study linked with the beginnings of competitive production, it is not always easy to calculate what the costs will be at the next stage of development. Many of those who are involved in non-ferrous die casting will recollect that after the pilot stage, the next one took place when the plant had a dozen or so machines all working well and with 'overheads' brought to a minimum. During this stage an attractive profit margin can sometimes be earned, although the total output is not large. The third stage comes when the plant must be expanded considerably to offer casting production to a world wide, as distinct from a limited, market and this brings many risks but sometimes satisfactory rewards. The production of ferrous die castings will possibly follow the same pattern.

References

1.   FRYE, G.R.
      Permanent mold process as applied to production of gray iron castings. *Modern Casting,* **54,** (4), 52 (October 1968)

2.   CARVER, B.G.; STREET, A.C.
      The Diecasting Book Portcullis Press Ltd., Redhill, UK p. 287

3.   SELLORS, R.G.R.; HEYES, J.G.
      Ferro-Di — GKN's new steel-diecasting process. *Foundry Trade Journal,* **141,** (3097), 811 (October 28, 1976) and **141,** (3099), 1024 (November 25, 1976)

4.   CROSS, R.
      Ferrous diecasting. *First South Pacific Congress,* Paper No. 80–27. Melbourne, 1980. Society of Diecasting Engineers, Australia

Chapter 18

Automatic metal transfer systems

With the stimulus of incentives, competent manual die casters of 30 years ago were capable of producing small aluminium die castings at speeds comparable with those of modern automatic machines. They had to operate the opening and closing mechanism, core withdrawal, and die lubrication as well as metal ladling. However, during the 1950s it became evident that automatic metal transfer systems must be developed for cold chamber machines. This was partly because some operators found lighter employment elsewhere in more pleasant conditions than those of the old fashioned foundry. At the same time the scope for pressure die casting was being widened so that large castings could be made, requiring such heavy ladlefulls of molten metal that manual labour could not cope unaided. It was not difficult to design cold chamber machines which would open and close the dies automatically but the automation of metal pouring presented many problems. The development of the processes for molten metal transfer proceeded in three stages.

At first the ladles were supported by some form of arm so that the operator could swing the ladle towards the pouring position without having to support the full weight of the metal. Sometimes a large ladle capable of holding the metal was stationed at the top of a length of channel; the operator would fill it with two or more lighter ladles and then it was tilted so that molten metal ran down the channel into the injection sleeve. The next stage, characteristic of the 1950s, involved the development of many different systems[1], including efforts to reproduce the movement of the manually operated ladle and various methods of forcing molten metal to the injection sleeve by air pressure. The prototype metal transfer devices rarely gave the same metal quality as the manual die caster had been able to achieve. Air pressure systems were difficult to control well enough to guarantee the same amount of metal for each shot, while early mechanized ladles did not pour in the controlled manner of the experienced manual operator. In some of these methods a skull formed on the ladle and this oxidized metal would subsequently be injected into the die. There were other problems because the linking of die opening and closing with the position of the ladle was not always regulated correctly.

During this period experiments were also made in exotic materials which, even today, have not been used successfully in the automatic transfer systems. For example, it was an attractive prospect that if the hot chamber process could be applied to the die casting of aluminium alloys, there would be great opportunities for increased rates of production. The hot chamber system had been used for many years with zinc alloys and, during the 1960s, hot chamber magnesium die casting was developed. Unfortunately the dissolving effect of aluminium on ferrous materials prevented conventional hot chamber machines from being used, so experiments were made with ceramic materials. However, as will be discussed later, they have not yet been able to provide the necessary robustness to endure the conditions of a die casting production department.

The third stage of development, during the 1970s, was the gradual elimination of the various processes which had failed to give the necessary high quality of metal and reliability. Many companies found that too much maintenance and supervision was needed to obtain satisfactory and economic production, with equipment that did not measure up to the requirements. Consequently painful decisions had to be made to discontinue the use of expensive but unreliable systems, and change to other types which performed more satisfactorily.

The metal metering equipment that has been tested[2,3] in the past 30 years can be grouped into six basic principles and there are several options available within some of them.

1. Air pressure systems   — with airtight furnace
                                         — with airtight crucible.

2. Mechanical immersed pumps.
3. Mechanical transfer systems   — ladle open or closed
                                         — movement rotary or 'up and over'
                                         — pneumatic, hydraulic or electric drive.
4. Electromagnetic feed systems   — with pump
                                         — with launder.
5. Vacuum feed systems.
6. Various experimental systems.

Air pressure systems

Figure 18.1 illustrates two versions of metering systems, where the furnace has a pressure-tight cover; by pressurizing with air or an inert gas, liquid metal is discharged through a delivery spout into the shot sleeve. For most aluminium and copper alloys, conventional ceramics are suitable construction materials. Accuracy in the amount of metal delivered at each cycle is controlled by the gas pressure-time relationship. Since this will be dependent on the amount of metal within the furnace at any given time, methods were developed to measure metal levels and automatically control the pressure-time sequence. These included ultrasonic and capacitance devices, low voltage metal probes, immersed bubble tubes and feedback correction for automatic measurement of the quantity of metal poured into the shot sleeve. The attractions of these systems include freedom from moving parts and speed of operation but, because the applied pressure must be related to the amount of metal in the crucible, the accurate working of the regulating devices is of great importance. It was apparent that, where an expert and dedicated maintenance engineer was in attendance, air-operated systems worked satisfactorily but if they were left without careful supervision, breakdowns and irregular metal deliveries resulted.

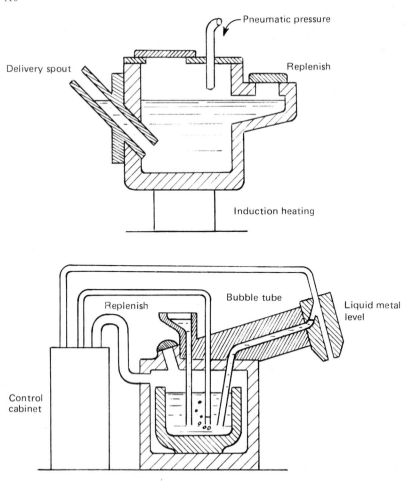

Figure 18.1 Two types of pressurized metal metering furnaces.

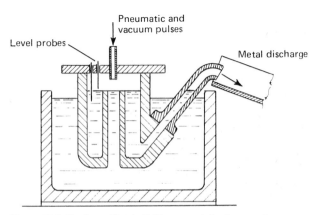

Figure 18.2 Lindberg 'Autoladle' immersed displacement pump

Methods were introduced which made possible the use of existing bale-out furnaces with the addition of airtight crucibles and immersion displacement pumps. A typical design showing the probes to control the amount of metal metered is illustrated in *Figure 18.2*. Filling of the crucible or chamber is achieved by pressure reduction while metal dispensing is controlled by increasing the internal pressure.

Mechanical immersed pumps

Hot chamber machines for zinc alloys, and sometimes for magnesium, use immersed direct acting pumps, but such methods are not suitable for aluminium, since it has a high affinity for ferrous metals. Research into alternative materials for constructing pumps to dispense aluminium alloys has involved the testing of ceramics such as silicon nitride, titanium diboride and zirconium diboride. While these were shown to fulfil technical requirements for hot chamber aluminium die casting, failures due to brittleness, with consequent high replacement costs, precluded their commercial use. In a series of works tests in the USA the reduction of production costs due to the faster output of the hot chamber machine was compared with the cost of titanium diboride goosenecks and pumps. Under ideal conditions the ceramic material came near to being justified but the equipment being tested was not sufficiently robust to ensure long life; in some tests the ceramic pumps broke down after a short length of service and the tests were discontinued, temporarily.

In Britain and other countries increasing use is being made of silicon nitride[4] as riser feed tubes in low pressure die casting machines. It is produced by reacting silicon with nitrogen at 1450°C. There are several forming methods ranging from the isostatic pressing of silicon powder into billets, casting of a powder slurry or pressing and injecting the

Figure 18.3 Applications of silicon nitride. (Courtesy Advanced Materials Engineering Ltd.)

powder and a binder into a die. After partial nitridation[5], the parts are machined with hardened tip tools and the final shapes completely nitrided in controlled atmosphere furnaces.

Figure 18.3 shows several applications of silicon nitride, including thermocouple protection sheaths which represent one of the earliest ancillary uses in die casting. A variety of seats and plugs are illustrated, similar to those which are used within the metallic crucible of the 'Telemetal' dosing system mentioned on page 76.

Although not yet commercially successful, development of a complete ceramic bale— out ladle has achieved long life, while the metal pours quickly and cleanly, leaving the ladle free from dross. Pumps constructed of ferrous materials, shown in *Figure 18.4,*

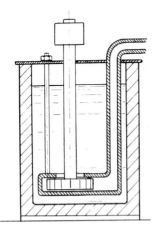

Figure 18.4 Metal Pumping Services immersed impeller pump

have been used for dispensing zinc and magnesium which do not attack the ferrous metal to any serious extent. For aluminium the impeller needs to be made of a refractory material but such components are liable to breakage, particularly when aluminium oxide build-up is formed on the impeller. Accurate control of the amount of metal dispensed is difficult to achieve unless the impeller is left running and metal delivery is metered with a stopper valve in the discharge tube.

Mechanical transfer ladles

Mechanical ladles are becoming the most popular method of transferring metal to small and large cold chamber machines, due to their relative cheapness and simplicity of operation. *Figure 18.5* illustrates a mechanical ladle, familiarly known as a 'Dipping Duck' using a tube of refractory-coated mild steel, connected to a pivot arm for extended reach and direct pour into the shot sleeve. The tube tends to pick up oxide from the surface of the melt; due to the length over which molten alloy is dispensed, turbulence and loss of temperature can be troublesome. To maintain the entire tube length at a sufficiently high temperature and avoid excessive metal oxide build—up, additional heating is required. This is usually provided by means of gas jets, positioned under the tube.

Other systems employ ladles which are either connected to a rotary pivot arm *(Figure 18.6)* or, where extended reach is required on larger die casting machines, use is made of

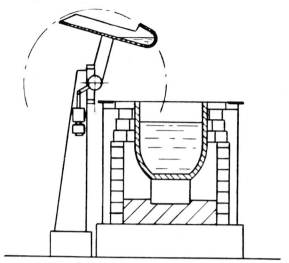

Figure 18.5 Mechanical ladle of 'dipping duck' type

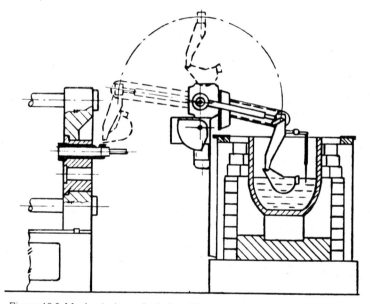

Figure 18.6 Mechanical transfer ladle with rotary movement

an overhead beam *(Figure 18.7)*. Ladle bowls are often cast in grey iron containing about 1% phosphorus, and offer satisfactory resistance to attack by molten aluminium, when protected with a refractory wash. Zirconium oxide pastes have been found effective in delaying the adhesion of molten alloy to the ladle by up to several days, although the normal practice is to coat ladles daily. Metal can be poured directly into the shot sleeve without the need for extended launders and, since the ladle is returned to the melt after each shot, supplementary heating is not required. When the ladle is dipped into the crucible, surface oxide is not entrapped because the metal flows into the ladle inlet underneath the bath level. Low voltage metal level sensor probes attached to the ladle

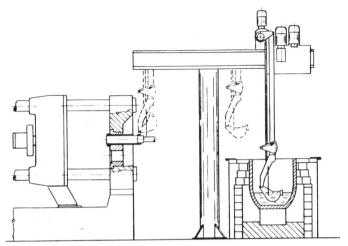

Figure 18.7 Mechanical transfer ladle with up-and-over movement

bowl act as safety stops to control the depth to which the bowl enters the melt. Maintaining a constant level of melt in the furnace is not therefore of importance.

Mechanical valved transfer ladles, made in Switzerland, achieve dross-free filling and discharge by employing a bottom valved crucible[6] activated by mechanical stops. As with other transfer ladles using metal sensor probes, varying metal levels in the crucible

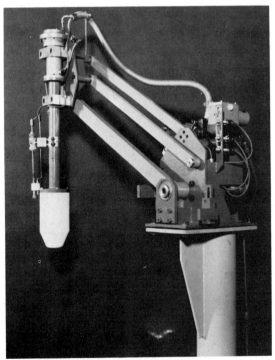

Figure 18.8 Mechanical valved 'Fillmat' transfer ladle.
(Courtesy Buhler Brothers Ltd.)

Figure 18.9 Hodler 'Telemetal' transfer system with cold chamber die casting machine. (Courtesy Bühler Brothers Ltd.)

can be accommodated but, without an overspill facility, the volume of metal collected relies on the probes, which must be maintained clean for the most effective control. *Figure 18.8* shows a ladle connected to a rotary pivot arm. *Figure 18.9* shows a 'Telemetal' ladle travelling on an overhead beam.

Electromagnetic transfer of liquid metal

During recent years, several types of metal pumps using the principle of electro-magnetic propulsion have been developed[7], with applications ranging from pumping liquid sodium in nuclear fast breeder reactors to the automatic transfer of aluminium in sand, low pressure and permanent mould casting. At present electromagnetic pumps are being tested and used for production in several pressure die casting plants.

A linear motor, operated by electric current and magnetic field, causes molten metal in a ceramic tube to flow and thus the metal supply can be directed to the shot cylinder of a die casting machine. *Figure 18.10* shows the working arrangements of the Novatome pump, developed in France. A ceramic base, about 250 mm in diameter, attached below the linear motor, is immersed in the bale-out furnace so that molten metal can be pumped through a vertical channel, from the bottom of the base, continuing upwards at an angle of about 10 degrees, with a downturn at the end directly above the shot sleeve of the die casting machine. In a typical pump, about 1 kg of metal can be delivered per second. Allowing the time of about 2 seconds for each cycle to commence and 1–2 seconds after delivery of the metal, it will be seen that such a pump is suitable for the manufacture of medium sized die castings with a production rate of the order of 100–150 shots/hour.

The power is obtained from mains electricity and no additional transformers or other ancillary equipment are required. During the brief time that the pump is trans-ferring molten metal, about 12 kW are used. The ceramic base is immersed in the molten metal and extracts metal from below the surface; in addition, the downturn of the delivery tube is near the injection sleeve, so oxide contamination does not occur. Since pressure die casting bale-out furnaces are replenished intermittently the metal level drops during the course of production and rises again as more metal is brought to the furnace. The electromagnetic pump is in a fixed position and, as the metal level drops, the electric power has to be adjusted automatically to maintain the required rate of metal transfer. This is accomplished by a monitoring device which signals the level of metal to the pump control cabinet. In order to obtain the highest quality of metal transferred to the injection cylinder, the upward inclination of the delivery tube should be about 10 degrees. This allows the pump to control the movement of molten metal, restricting gravity flow to the end of the downturn just above the shot sleeve. When the pump is not operating, the delivery tube remains full of liquid metal held at the correct temperature by electrical heaters surrounding the tube.

Bale-out furnaces with only small openings are not suitable for the installation of present designs of electromagnetic pumps; a furnace with a well for receipt of molten metal from the travelling ladle and another for positioning of the pump gives the best working conditions. Melter-holder furnaces such as the immersed crucible furnace are also suitable. The cost of electromagnetic pumps is comparable with mechanical ladling devices. In addition to the undoubted high quality advantages, the electromagnetic pump

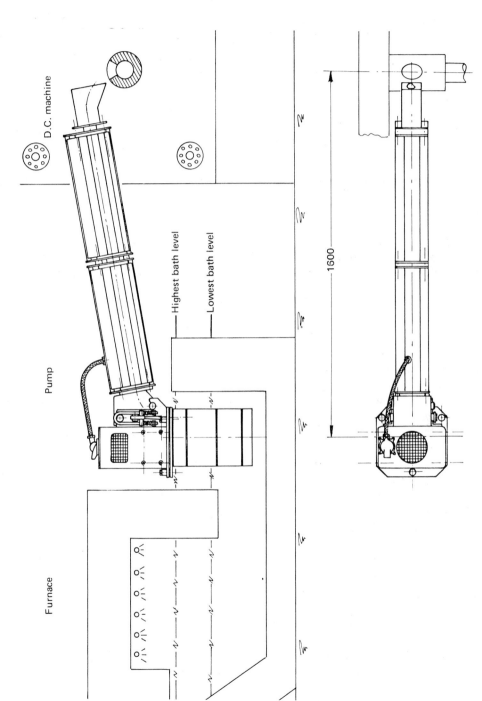

Figure 18.10 Electromagnetic metal pump. (Courtesy Novatome Liquid Metal Systems)

has no mechanical working parts; maintenance does not involve any sophisticated methods and the system is adaptable to all metals and alloys. Pumps capable of metal transfer up to 3.5 kg/second are also available, but they are about 350 mm in diameter, and are more suited for die casting machines equipped with large furnaces.

Vacuum feed systems

Vacuum filling is a displacement process but, by the use of a negative pressure in the die casting machine injection sleeve, atmospheric pressure on the free metal surface forces molten alloy into the feed tube. *Figure 18.11* illustrates the principles of control and it will be noted that special design of dies and injection equipment are necessary to accept the technique. The system has proved successful with die casting machines having vertically reciprocating platens, and extra benefits have been reported from evacuation of the die which in turn has led to improved casting soundness discussed on page 232. Other attractions are freedom from moving parts, metal supply taken from beneath the melt surface is dross-free, and the entire furnace can be enclosed for improved environment and reduced energy costs.

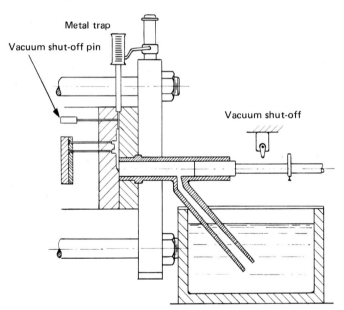

Figure 18.11 Westinghouse vacuum delivery system

Other ladling methods

1. A development recently exploited for die casting ferrous alloys employs induction melting of solid slugs; details were given in Chapter 17. The system has been considered for other alloys but due to their lower electrical impedence, speed of melting is low and energy costs are high.

2. Robots could be utilized for the ladling of metal, although due to their relative sophistication they are particularly suitable for casting retrieval and other ancillary operations which often need to be reprogrammed and combined in one casting cycle. It is possible for a robot to service two machines, although the less sophisticated mechanical ladling systems are being used successfully and with a lower cost than robots.

3. Fluidic devices using air have been employed in process control equipment and it was natural that research efforts be directed to metal metering. Water models have already confirmed suitability of the principle for denser fluids and development into die casting could well incorporate a heated circulating tube positioned over the machine pouring spout and returning to the molten alloy in the furnace. A fluid valve adjacent to the pouring spout could direct a metered amount of alloy into the die casting machine.

4. The die casting of alloys by 'Rheocasting' and 'Thixocasting' techniques is still at an experimental stage[8,9] but successful casting of aluminium alloys by these techniques is reported on a pilot plant scale. These processes are attractive for die casting of copper and ferrous alloys and, should the continued researches be successful, they can be expected to lead to significant advances in technology, including the die casting of aluminium alloys. These developments are discussed in Chapter 24.

References

1.   HILL, T.B.
        Liquid Metal Transfer: survey of existing methods of automatic metering. *BNF Metals Technology Centre Report No. 1852* (October 1973)

2.   (Editorial)
        Automatic ladling in non—ferrous foundries. *Diecasting and Metal Moulding,* **8,** (9), 28 (May/June 1978)

3.   HILL, T.B.; STREET, A.C.
        The Diecasting Book Chapter 25 Portcullis Press Ltd., Surrey and American Die Casting Institute, Des Plaines, Illinois (1977)

4.   FLYE, D.J.
        Silicon Nitride. Its uses in diecasting. *Diecasting and Metal Moulding,* **8,** (3), 7 (May/June 1977)

5.   TAYLOR, D.N.
        Silicon nitride: a material for the future. *Chartered Mechanical Engineer,* **25,** (7), 44 (July 1978)

6.   KOCH, P.
        Simple methods for stepping-up production *Buhler Brothers Technical Information Brochure No. 5.* Buhler Brothers, Switzerland

7.   PRITCHETT, T.R.
        Application of Electromagnetism to Aluminium Metallurgy. *Light Metal Age,* 31, (11/12), 21—24 (December 1973)

8.   FLEMINGS, M.C.; MEHRABIAN, R.
        Casting semi—solid metals. *Modern Casting,* **63,** (12), 31 (December 1973)

9.   JOLY, P.A. and MEHRABIAN, R.
        The rheology of a partially solid alloy. *Journal of Material's Science,* **11,** 1393—1418 (1976)

Chapter 19

Metal melting treatments

Forty years ago die casting foundrymen used simple fluxes to clean their metal; salt, borax, glass and charcoal were employed according to the preferences of supervision, often based on routines used in sand foundries. In those days, problems of health and the environment were often overlooked and clouds of fume were all too frequently seen and inhaled whilst the metal was being treated. Fortunately, a scientific approach to the subject was developing and today metal treatment is based on principles of good chemical and metallurgical procedures and the requirements of health, safety and clean air.

During the course of manufacture from primary metal to finished die castings, at least three metal melting stages occur where metal treatments can be employed to ensure freedom from oxides and other inclusions. First the metal is made either at a primary producer of zinc alloys, or at a secondary metal refiner for aluminium alloys. The latter subject has already been discussed on page 20 and it will have been evident that the treatment of aluminium scrap requires a significant amount of chemical additions to bring the metal to the required degree of cleanness and purity.

Next, the metal is supplied either as ingots which have to be melted in bulk melting furnaces, or melting/holding furnaces at the die casting machine, or it may be delivered in the molten condition and is then held in large containers, as discussed on page 51 prior to distribution through the die casting plant. Finally, the molten metal, having arrived at the die casting machine holding crucible, must be maintained in as clean a condition as possible.

Whatever the stage in the preparation of molten metal the first priority is to avoid overheating or any other circumstance which will cause oxide and impurities to accumulate. Covering fluxes or other additions protect the surface of the molten metal from contaminants, but the excessive use of such treatments must be regarded as a failure to maintain efficient metallurgical control. Lavish treatment with fluxes add to the cost of production and reduce the life of crucibles and refractories. The drosses so formed may appear to be worthless non-metallic residues, but may actually contain over 50% of metal as distinct from compounds of the metal — a waste of money that is not always apparent. Some problems of metal losses and the measurement of these losses are discussed in Chapter 20.

Fluxing

Fluxes are normally mixtures of salts which react with oxides and impurities either mechanically, chemically or both to form a dross which can be removed easily, leaving a clean melt surface. The main purpose of a flux is to protect the metal surface from oxidation which accelerates with increase in temperature and the time held within a furnace. In addition, fluxes assist in removing unwanted metallic or non-metallic impurities from the molten alloys by providing a substance with which these elements or compounds may combine. The ferrous and non-ferrous casting industries owe a great deal to the work of Dr K. Strauss on casting techniques and chemical treatments of metals and alloys. His book 'Applied Science in the Casting of Metals'[1] contains information on the manufacture and application of fluxes on a scientific basis.

Formulating a flux for a given alloy is a chemical problem requiring knowledge of the composition and properties of all materials involved. Salt mixtures, usually chlorides with low melting points, are the basis of many fluxes. Chloride salt mixtures do not appear to have any chemical action on the oxides but separate them from the base metal by mechanical means. Work by Sully, Hardy and Heal[2] on aluminium alloys indicated that the wetting properties of chlorides assist in releasing aluminium oxide by penetrating the interface between oxide and metal. More active fluxes are usually necessary to react chemically with metal oxides. Fluorides of calcium and sodium, which produce heat by exothermic reaction, have a dissolving action on the oxides and are added to basic chloride fluxes in concentrations up to about 25%. The function of these fluxes is believed to be firstly a reaction with the base metal at the oxide-metal interface which detaches the oxide film and removes it as a suspension into the dross layer. Molten fluxes, although effective in cleaning and protecting metal from oxidation, are difficult to remove from the melt surface and often contain a high concentration of entrapped metal. Improvements can be made to these chloride—fluoride flux mixtures by the addition of small quantities of oxidizing agents such as sulphates or nitrates. These generate heat in the dross layer by exothermic reactions and encourage fine globules of molten alloy to coalesce and fall back into the melt. The flux layer becomes dry and powdery, absorbs oxides and produces a dross which is easy to remove[3].

Fluxes are usually applied to the melt surface by hand while compressed air guns assist dispersion over the melt surface in large furnaces. Light rabbling of the flux into the surface layers of the melt assists the reaction, but deeper stirring must be avoided since it can lead to suspended particles being carried over into the castings.

Treatment of secondary aluminium alloys

Raw material scrap used by the secondary aluminium refiner ranges from large castings and sheet to fine turnings and aluminium foil, often contaminated with iron and other elements as well as oil, grease, paint and paper. In order to recover the maximum amount of usable metal, various methods are employed for sorting and treating scrap to remove unwanted metallic and non-metallic materials before remelting. Reverberatory furnaces are generally used for melting and refining large scrap while rotary furnaces are used for lighter contaminated scrap. All material is charged through a surface flux cover which can amount to 40% of furnace capacity, depending on the type of charge. Mixtures of calcium, sodium and potassium chlorides melting at about 500°C are suitable for large scrap while fluxes melting at about 600°C based on sodium and potassium chlorides and cryolite (sodium—aluminium fluoride) are suitable for lighter scrap. Aluminium scrap containing more magnesium than the specification of the alloy allows can be refined using a flux with a high proportion of sodium chloride or sodium fluorosilicate. Alternatively chlorine gas is bubbled through the melt to form magnesium chloride which will rise through the melt and assist in removing aluminium oxide, aluminium nitride and siliceous compounds by absorption.

During melting, water vapour reacts with aluminium to form aluminium oxide and hydrogen. In a similar way, nitrogen reacts with aluminium to form aluminium nitride which, together with aluminium oxide, accounts for the majority of impurities formed during the melting of aluminium alloys. Being practically insoluble in aluminium they remain as suspended particles within the melt, leading to inclusions in castings.

Metal treatments in the die casting plant

The simplest and cheapest flux that can be used to form a protective molten cover on the surface of aluminium consists of a 50/50 eutectic mixture of sodium and potassium chloride. This melts at 658°C and assists in separating the surface oxide film on the melt from the underlying aluminium. To encourage the formation of a dry powdery dross low in metallics, materials encouraging exothermic chemical reactions are added to the basic salt mixture in varying amounts up to about 25%. For this purpose, cryolite or sodium fluoride are used with oxidizing compounds of sulphates and nitrates.

Several reactions occur during the production of a powdery dross. Oxidizing agents react with particles of molten aluminium entrapped in the dross cover to produce alumina and in so doing increases the dross temperature. A second exothermic reaction is encouraged where aluminium particles react with fluorides to produce aluminium fluoride and the unstable gas aluminium subfluoride. This reacts with oxygen in the air to form aluminium oxide and aluminium fluoride which continues to take part in the reaction until all the fine aluminium particles are oxidized. The fluorides remove oxide films surrounding larger aluminium particles while heat generated during the exothermic reaction encourages these oxide free particles of aluminium to coalesce and fall back into the melt, leaving a dry powdery dross which is easily skimmed from the melt surface. A typical untreated aluminium dross contains about 80% aluminium and 20% aluminium oxide but with good fluxing practice about half of the metal can be recovered.

Aluminium—magnesium alloys require a different flux, since the alloys oxidize and absorb gas more readily than other aluminium die casting alloys. A flux with a low melting point, strong cleansing action and low moisture content is recommended where a protective fluid layer can be formed at about 440°C. These flux mixtures are normally made from various chlorides and fluorides with the major constituent being anhydrous magnesium chloride. It is essential that sodium compounds are not used, due to their deleterious effect on aluminium—magnesium die casting alloys. Even at concentrations of 0.001%, sodium has an embrittling effect and makes these alloys hot-short.

Zinc alloys

Fluxing is not normally carried out when melting pre-alloyed zinc ingots, but treatment is recommended where scrap returns are included with an ingot charge. Scrap is inevitably contaminated and can lead to the formation of oxide and the intermetallic compound $FeAl_3$ which floats to the top of the melt. Where a large amount of scrap is involved, it is good practice to treat it separately from the normal bulk melting system where thorough fluxing and composition adjustments can be made before further use.

Early fluxes using ammonium and zinc chlorides produced a fine powdery dross, allowing entrained oxides and intermetallics to be removed easily whilst a high percentage of entrapped zinc was released back into the melt. Unfortunately chlorides and magnesium have an extremely strong affinity for each other and chloride fluxes will reduce magnesium in the melt by volatilization of magnesium chloride. With chloride addition of more than about 1.5% of the melt weight, all the magnesium in the alloy would be removed. In addition, exothermic reactions generate copious amounts of fumes, creating pollution problems. Fluxes now in general use for refining zinc alloys contain inorganic fluorides and magnesium salts added to the basic ammonium and zinc chlorides. Fuming problems have been minimized and magnesium salts help to counteract loss of the small but essential magnesium content in the zinc alloy.

Magnesium alloys

Magnesium is a reactive metal and oxidizes readily; at temperatures above the melting point the metal can burn if precautions are not taken to prevent access of air. Unlike aluminium oxide, which protects the base metal from further oxidation, magnesium oxide is porous, allowing oxidation of the underlying melt to continue.

Bulk melting of magnesium alloy is often carried out in tilting furnaces. One of the earliest used fluid fluxes was the mineral Carnallite consisting of about 60% magnesium and 40% potassium chloride which melts at $380°C$. Magnesium chloride wets magnesium oxide and increases its density to encourage some settling to the bottom of the melt, but the fluid flux of similar density to magnesium can remain suspended in the surface layers of the melt and lead to flux inclusions in the castings[4]. Refining fluxes of stiffer consistency which absorb the fluid flux, oxides and inclusions are produced by adding compounds of magnesium oxide and calcium fluoride in concentrations up to about 35% to the protective flux mixture. Viscosity is maintained over a wide temperature range, avoiding dross being entangled in the metal and ensuring its easy removal from the melt surface.

While molten magnesium alloy is contained in the die casting machine holding furnace, continuous protection must be given to the melt surface with a protective cover of sulphur dioxide or sulphur hexafluoride as described on page 145. These gases, being heavier than air, fill the canopy enclosure and, by allowing a slight leakage of gas to occur, ingress of air is prevented. On hot chamber machines, lateral perforations on the gooseneck allow the plunger in its upper position to be surrounded by metal so that magnesium residue cannot accumulate.

Brass

Flux protection of brass is required, particularly in the secondary industry where zinc fume and dross are produced during alloy manufacture. Of the many flux compositions available, borax is a common constituent, with various combinations of either sodium chloride, sodium carbonate, sodium fluoride, fluorspar, cryolite and charcoal. Following evaluation of many fluxes[5] it was found that, in general, dross residues after fluxing operations contained up to 50% zinc and 20% copper.

The need to use fluxes when melting and holding brass for die casting is often eliminated when clean metal is used. When necessary, fluxes can assist in separating non-metallics from the melt and suppress the evolution of zinc vapour.

Degassing

Although it is usually unnecessary to degas melts for pressure die casting, it may be used for aluminium alloys, particularly where castings that must be pressure tight undergo a substantial amount of machining beyond the dense surface skin. In addition, degassing will assist in bringing suspended oxides to the surface of the melt where they can be removed with the surface dross.

Hydrogen is the only gas normally in contact with die casting alloys which has any degree of solubility. Reaction between molten metal and water vapour produces hydrogen in its atomic state which can dissolve in the melt, although molecular hydrogen has a limited solubility. Water vapour is introduced through a variety of sources including humid atmospheres and combustion of fuels. The extent to which dissolved hydrogen can lead to microporosity depends on the solubility of hydrogen in the metal as it solidifies. There is shortage of information on the extent of hydrogen solubility in die casting alloys, but solubility is dependent on pressure and the temperature and composition of the alloy. In addition, the aluminium oxide skin formed on aluminium and its alloys gives some protection against gas absorption. Work by Ransley and Neufeld[6] on the solubility of hydrogen in pure aluminium has shown that at the liquidus point, about 660°C, 100g of aluminium can contain up to 0.69 cm^3 of gas, while its solid solubility just below the liquidus is reduced to 0.036 cm^3. At 300°C the gas solid solubility is reduced further to 0.001 cm^3. If molten aluminium contains appreciable amounts of gas the majority of it will be trapped as the metal is rapidly cooled, leading to the commonly known 'pin hole' microporosity. Talbot and Granger[7] have shown that the small amount of residual hydrogen is not entirely in solid solution in the aluminium, but is distributed between solid solution and a dispersed internal gas phase. The rejected gas can be nucleated both in the liquid during solidification and in the solid immediately afterwards, thus generating two different kinds of porosity, distinguished by the terms interdendritic, or primary, porosity, and secondary porosity, taking the form of evenly distributed spherical voids.

In contrast to aluminium, the solid solubility of hydrogen in magnesium is several hundred times greater, as illustrated in *Figure 19.1,* and therefore, the chances of magnesium die casting alloys having microporosity is reduced. Talbot[8] concludes that hydrogen evolution and solidification contraction of castings contribute to porosity, but for magnesium the contraction is the dominant factor. This is due to the lower decrease in hydrogen solubility on solidification, so that a hydrogen content in the residual liquid sufficient to

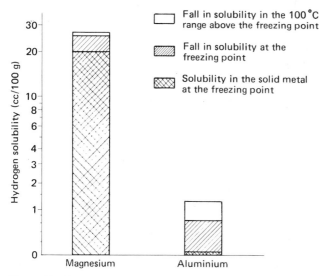

Figure 19.1 Solubility of hydrogen in magnesium and aluminium

nucleate gas bubbles is not reached until a later stage in solidification, if at all. If the hydrogen content is low, the total volume of porosity is determined by the contraction of the metal. Alternatively, if the hydrogen content is so high that the volume of gas released is greater than the volume change due to contraction of the metal, the volume and distribution of porosity are both controlled by the dissolved gas. Zinc does not absorb gases to any extent so degassing, which would also remove magnesium from the melt, is unnecessary. With brass, the zinc vapour pressure is sufficient to prevent the ingress of hydrogen.

Hydrogen can be reduced from melts by treatments with other gases or chlorides at the bulk melting furnace or the die casting machine holding furnace. Both methods rely upon a steady stream of gas rising through the melt into which dissolved hydrogen diffuses and then escapes into the atmosphere. This diffusion continues until the partial pressure in the gas bubbles and in the metal are in equilibrium, after which only a small amount of hydrogen will be left in solution.

Methods include use of the inert gases nitrogen, helium and argon or the active gases chlorine and fluorine. Treatment times are much longer with inert gases and there is the possibility that at temperatures over 700°C in the treatment of aluminium alloys, nitrogen can form aluminium nitride, or magnesium nitride in alloys containing magnesium. These compounds are practically insoluble in aluminium and will remain suspended in the melt. Chlorine reacts with molten aluminium to produce aluminium chloride which is volatile above 180°C and bubbles through the melt encouraging hydrogen diffusion. Although this technique is more efficient than nitrogen degassing, there are toxicity hazards, so efficient ventilation systems must be used. In addition, there is the loss of aluminium, and in alloys containing magnesium, this element is lost as magnesium chloride. Chlorine can be diluted with nitrogen to reduce these problems but with some decrease in degassing efficiency.

Plunging metal chlorides into the melt is an alternative technique and, following a reaction with molten aluminium, aluminium chloride is evolved which bubbles through the melt. Careful selection of these chlorides is necessary since some of the metals are often unwanted impurities in die casting alloys. In addition, most metal chlorides absorb moisture and there is the risk of introducing more hydrogen than is taken out. The most popular material used today for degassing is hexachlorethane, C_2Cl_6. This crystalline material does not absorb moisture or introduce unwanted constituents into the melt, and has a degassing efficiency comparable to that of chlorine. It is added to molten aluminium in tablets and forms aluminium chloride, chlorine and carbon tetrachloride which bubble through the melt. Other incorporated salts delay the rate of decomposition and encourage a more thorough gas purging by producing smaller bubbles.

Gaseous treatments are also effective for removing relatively large inclusions, but for small inclusions it is necessary to use the previously described fluxing treatments. A recent report by Emley and Subramian[9] compared the melt treatment processes that have been developed, many of which are suited to the refining of aluminium and combine degassing with the removal of inclusions. A paper by Japanese researchers[10] compares several processes of this type which use in-line filters of ceramic, glass, coke, graphite or other materials. Usually a large filter surface area is provided through which the metal in the bulk melting furnace flows, and mixtures of argon and chlorine can be passed through the filter layer so that simultaneous degassing can be carried out.

Fume collection and treatment

Many die casters use collection hoods to capture exhaust furnace gases and fumes emitted when holding and treating molten metals. Extraction can be either directly to atmosphere or via a cleaning plant. Emissions do not always constitute a health hazard but they are a nuisance, and in many countries they are subject to control. When collecting and cleaning systems are being considered, it is important to understand the many problems that can arise due to high temperatures, the amount of particles that can settle in a duct and fan system, and the corrosive effect of fume products. To reduce the corrosion problems, use of rubber, plastic-lined steel, stainless steel and reinforced polyester plastic can be specified, although it is necessary to cool the fumes to avoid damaging some construction materials. Before consideration can be given to fume collection and treatment, information is required of the volume of effluent gases, the particle size distribution and composition and whether the emissions are continuous or limited to certain parts of the operating cycle.

When melting aluminium alloys, the major source of fume arises when treating the melt with fluxes normally based on sodium chloride and fluoride salts. Mantle and Garner[11] gave results of emissions when refining secondary aluminium alloy in rotary furnaces where low grade materials are usually handled and large quantities of flux are consumed (*see* page 21). Over 70% of the solids were reported to be sodium chloride; this high level is probably due to unreacted flux particles carried within the products of combustion. The fumes do not constitute a health hazard but are strongly hygroscopic and form a white fog in moist air. Where chlorine is used to reduce the magnesium content in a melt, below about 0.2%, increasing amounts of aluminium fluoride form and hydrolyse to produce alumina and hydrochloric acid fume.

The highest efficiency collection plant is required for handling the fume and submicron sized particles emitted. Fabric filters, wet scrubbers and electrostatic precipitators have been used, although, at present, fabric filters seem to be the most favoured and are claimed to be operating successfully in several plants. When a material resistant to acid attack is required, good performance has been obtained using polyester felt, but, since few fabrics are suitable for continuous operation above 200°C, the operating temperature range for a fabric filter is restricted. Forced draught coolers are often used to reduce the possibility of hot carbon particles and other burning debris damaging fabric filters.

Problems can be met due to the corrosive nature of fume when its temperature falls below dewpoint. A particular example of this was associated with an aluminium bulk melting furnace in a die casting plant where a scrubber containing over 5000ℓ of water showed severe corrosive attack of certain bitumen-coated mild steel sections. After less than a month's operation, about 18 kg of iron was corroded away and the water solution had an acid pH level of 2.6. Analysis of the solution confirmed the iron loss and gave the results shown in *Table 19.1*.

TABLE 19.1

Contaminant	g/ℓ
Iron	3.4
Aluminium	0.06
Chlorides	1.0
Sulphates	6.9
Fluorides	Less than 0.30.

The choride content of the solution probably originated from the absorbtion of flux particles rather than the formation of volatile chlorides. The sulphate content was surprisingly high and must have come from the fuel oil used in the furnace. Approximate analysis of the flux used showing absence of sulphate was determined as:

Sodium chloride	70.0%
Sodium silicofluoride	15.5%
Sodium carbonate	10.0%
Calcium hydroxide	4.5%

Wet collectors have the advantage over other systems in being able to collect chlorine and hydrochloric acid fume but many precautions need to be taken to reduce acid attack including use of corrosion resistant materials and alkaline additions to the water.

References

1. STRAUSS, K.
    Applied science in the casting of metals. Pergamon Press, London (1970)

2. SULLY, A.H.; HARDY, H.K.; HEAL, T.J.
    An investigation of thickening and metal entrapment in a light alloy melting flux. *Journal of the Institute of Metals,* **82,** 49 (1953/54)

3.   KISSLING, R.J.; WALLACE, J.E.
      Fluxing to remove oxide from aluminium alloys. *Foundry,* **91,** (3), 76 (March 1963) (Many
      useful references are appended)

4.   EMLEY, E.F.
      Non-metallic inclusions in magnesium-base alloys and the flux refining process. *Journal of the
      Institute of Metals,* **75,** 431 (1949)

5.   DAWSON, P.R.; WILKINSON, A.
      Fluxing for the secondary brass industry. *The Metallurgist and Materials Technologist,* **11,** (1),
      35 (January 1979)

6.   RANSLEY, C.E.; NEUFELD, J.H.
      The solubility of hydrogen in liquid and solid aluminium. *Journal of the Institute of Metals,* **74,**
      599 (1948)

7.   TALBOT, D.E.J.; GRANGER, D.A.
      Secondary hydrogen porosity in aluminium. *Journal of the Institute of Metals,* **92,** 290 (1963/
      64)

8.   TALBOT, D.E.J.
      Effect of hydrogen in aluminium, magnesium, copper and their alloys. *International Metal-
      lurgical Review,* **20,** (1975)

9.   EMLEY, E.E.; SUBRAMIAN, V.
      In-line treatment of liquid aluminium by the FILD and other processes. *Metallurgical Society of
      the American Institute of Mechanical Engineers,* TMS paper, No A 74/62 (1974)

10.   INUMARU, S.; SHIROTANI, M.
      Recent developments in treatment of molten aluminium. *Sumitomo Light Metal Technical
      report,* pp.30–37. ISSN 0039–4963 (July 1978)

11.   MANTLE, E.C.; GARNER, A.V.
      Work on measurement and control of fume. *BNF Conference on Air Pollution.* BNF Metals
      Technology Centre, Wantage, Oxfordshire. (October 1969)

Chapter 20

The measurement of metal losses

Since raw material in the average aluminium alloy die casting accounts for about a third of the total cost, it is important to discover how much of this valuable material is lost in the foundry and machining shops and then determine how the metal loss can be reduced. The design stage for a component presents an early opportunity to exploit the possibilities of the process and minimize the amount of metal to be removed by subsequent machining. Components can be cast at minimum section, while obtaining the required strength by including webs. Coring of recesses can provide lightening.

During the past decade, the development of thin-wall zinc alloy die castings, discussed on page 103 has done a great deal to save material in that branch of the industry. A similar effort has been made in aluminium die casting with the 'cast-to-size' concept. The conditions of manufacture are controlled precisely, to provide a close approach to the finished product, thus eliminating some machining operations which had been thought inevitable before. Lower casting weights lead to less raw material cost, less metal to be melted and, therefore less melting losses. Furthermore since machining processes are reduced, the finishing costs are lower.

In addition to the possibilities of component weight reduction, there is considerable scope for reducing the weight of runner systems and overflows, provided that production rate and quality of the castings are not prejudiced. The following example shows how a modest saving was made by altering a double impression die that had been making a pair of aluminium components, dating back to the time when runner systems were designed without much concern for the large amount of surplus metal involved. The two castings weighed 0.2 kg each, while the weight of runner and overflows amounted to as much as 1.1 kg. The runner system was altered to reduce its weight to 0.45 kg — still slightly more than the weight of two castings. These die alterations led to a saving of metal melting and metal losses of £200 for a run of 10 000 castings. At that time the new technology for computer aided die design, discussed on page 231 had not been perfected. If the die had been designed with modern methods the metal saving would have been much more substantial.

Molten metal exposed to the atmosphere tends to oxidize; for example aluminium oxide Al_2O_3 is formed on the surface of aluminium, 1 kg of the metal being converted to 1.9 kg of oxide, according to the atomic weights of the two elements. This is a source of wastage which can be reduced by good housekeeping and melting efficiency, but a small loss is inevitable. Molten alloys of aluminium and other metals used in die casting are often treated with fluxing salts, as was discussed in Chapter 19. The flux partly lessens the likelihood of further oxidation, but it also releases included metallic particles so that as much as possible of the dross is oxide. Nevertheless, such drosses do contain metal, often a surprisingly large amount; if they contained only oxide they would be of little interest to refiners. Any company which is receiving a good price for their drosses should have them analysed and then consider how they can reduce the metal content of the drosses, even if that involves obtaining a lower price for them.

Some years ago the Authors were involved in an exercise to discover how much metal was entangled in drosses produced by two die casting departments, in which metallurgical control had not been very efficient. In one it was found that zinc alloy dross contained as much as 80% metal and aluminium dross 75%. In another section, aluminium alloy drosses contained about 50% reclaimable metal. It was realised that the higher metal losses occurred where the foundry personnel did not allow sufficient time for the fluxing salts to react and they had been skimming what they thought was only dross from melting furnaces without realizing the amount of metal that was being removed at the same time.

Now that metal costs have risen steeply, some die casting managements have been investigating methods of calculating their true annual metal losses. When it is considered that sometimes as much as 2 tonnes of aluminium has to be melted initially to produce 1 tonne of die castings, a 1% saving in melting loss can almost cover the cost of fuel for melting the metal. Melting loss can indeed be measured by melting a given weight of metal in a crucible and either measuring the weight of castings produced or weighing the dross that is removed under these conditions. Losses of 2–3% for aluminium, less for zinc and more for copper alloys are indicated. However, such losses are only part of the total; the discovery of the true loss per annum involves some quite complex 'metal accountancy'.

Metal balance sheets

In theory it should be no more difficult to make a metal balance sheet than a profit and loss account. It is necessary to state the weight of metal in stock at the beginning of the year, and add to it the weight of metal received during that year. Next the total weight of casting consignments is added to the weight of metal in stock at the year end. The difference between the two totals indicates the amount of metal lost. That can then be compared with the weight of castings sold, to provide an estimate of metal loss per cent, which is the correct figure to use in preparing price estimates and for the purpose of cost analyses.

When one begins to investigate the metal loss exercise, several problems come to light and it is necessary to consider them separately to determine how the individual die casting management can establish a sensible and appropriate method for metal loss measurement.

The total weight of metal in stock at the beginning of the period comprises the following:

1. Ingots in store and in the plant
2. Weight of metal in furnaces

3. Weight of castings in progress, either with runners attached or in a semi-finished state
4. Weight of finished castings not yet despatched
5. Weight of scrap and runners
6. Weight of plated or other scrap sent to metal refiners for treatment
7. Weight of customer rejects held in bond while causes of rejection are being investigated
8. Weight of castings containing inserts minus the weight of the inserts.

The total receipts of metal during the period comprises:

1. Weight of ingots received
2. Weight of molten metal received
3. Weight of rejected castings returned from customers
4. Weight of castings made elsewhere or made in the plant and sent outside for machining or finishing.

The total outgoings during a period comprise:

1. Nett weight of casting consignments calculated from the gross weight minus the weight of cartons, boxes, pallets or other forms of packing
2. Weight of castings sent to other branches of the same company
3. Weight of castings sent to outside contractors for machining or finishing
4. Weight of ingots or molten metal returned to suppliers for quality or other reasons.

The total weight of metal in stock at the end of the period will be measured under the same headings that were used for the estimation of opening stock.

Before the exercise is commenced it is desirable to assess the importance of each of the factors listed above. The aim will be to ascertain the extent of metal losses correct to about ½%; that is to say a final figure of 9% would be taken as an acceptable approximation between 8½ and 9½%. If an intelligent guess of any of the items will not prejudice the accuracy of the final figure by more than ¼% it is sensible to make such an estimate. For example it would be more trouble that it was worth to weigh the contents of each furnace at the time of the exercise. An experienced person from the Works Manager's or Metallurgist's departments should be able to assess the contents of each furnace with reasonable accuracy. As another example, if the total weight of inserts in castings during the year amounted to only 1 tonne out of a weight of 5000 tonnes of die castings sent out, an insert weighing exercise could be avoided.

One of the largest sources of inaccuracies occurs because most consignments are weighed in packages of varying size and weight. Some means have to be worked out for measuring the total weight of packing material and deducting it from the gross weight of consignments. This is difficult enough if only a few sizes of packages are used but some consignments have to be sent out with special wrapping, or spacers in the cartons, all of which introduce difficulties in establishing the nett weight. The other alternative is to have an accurate and up to date record of the weight of each casting. The estimated weights will not be accurate enough as most castings are different in weight from the first estimate. Also from time to time modifications are carried out, involving weight changes; the dies wear, and the castings become slightly heavier. The decision about how to go about this problem can only be taken after assessing how much valuable information will come from the exercise. One enterprising but meticulous part-time 'weight accountant' might be sufficient for keeping a daily record of all consignments, with a weight check on each type of casting. The Authors have discussed this problem in many areas during recent years and have not yet seen completely satisfactory 'metal accounting'. Yet companies are often willing to engage more cost or financial accountants and one wonders whether a similar enterprise in the metal loss area would give equally valuable results.

The following shows a metal balance sheet, giving figures typical of a medium sized die casting establishment.

A metal balance sheet

Stock at beginning of the period	Tonnes
1. Metal and ingots in stock	150
2. Estimated amount of metal in furnaces	34
3. Castings in progress	90
4. Finished castings not yet despatched	40
5/6. Scrap and runners	50
7. Rejects in bond	1
8. Castings with inserts minus weight of inserts	1
Total opening stock	366

Total intake

1. Weight of ingots received	1800
2. Weight of molten metal received	1700
3. Rejects returned	70
4. Subcontractors' deliveries	30
Total intake	3600
Plus opening stock	366
Intake plus stock	3966

Consignments during the period (outgoings)

1. Nett weight of casting consignments	3244
2. Weight sent to other branches	40
3. Weight consigned to subcontractors	20
4. Metal returned to suppliers	10
Total outgoings	3314

Stock at end of period	Tonnes
1. Metal and ingots in stock	170
2. Estimated amount of metal in furnaces	28
3. Castings in progress	77
4. Finished castings not yet despatched	60
5/6. Scrap and runners	35
7. Rejects in bond	2
8. Castings with inserts minus weight of inserts	1
Total stock	373
Plus outgoings	3314
Outgoings plus closing stock	3687
Compare with opening stock plus intake	3966
Loss of metal during the year	279

Comparing the loss of metal with the weight of casting consignments to customers a metal loss of about 8½% is indicated. It is a useful exercise to compare the calculated metal loss with the amount of dross sent out remembering that the weight of dross is only part of the total metal loss and that 'dross' consists of a mixture of metallic particles, oxide and other compounds.

Chapter 21

The metallurgy of die casting machines

Die casting machines built between the late 19th century and the First World War were simply mechanisms for injecting a metal of low melting point into a die, to produce small or medium sized castings of good surface but without much concern for their solidity. No great stresses were set up and the machines could be made of cast iron and mild steel. By the time of Doehler's improvements of die casting machines and the extension of the process to the manufacture of more highly stressed components, the machines had to be more powerfully constructed than before. The locking of the die and the injection of molten metal were done mechanically, or by the use of hydraulic force or air pressure. A Doehler machine of the 1920s contained eight ferrous metals and one of bronze. The following is a list of the major components. Some of the materials were defined with specification numbers and so can be identified. Others were given more general descriptions.

Shot cylinder	medium carbon cast steel
Pedestal	cast iron
Crosshead	0.4% carbon 1.0% chromium steel
Closing cylinder	medium carbon cast steel
Tie bars	0.04% carbon, 1.0% chromium, 0.2% molybdenum steel heat treated
Link pins	0.15% carbon, 0.7% chromium steel
Link bushing	bronze
Toggle links	0.4% carbon, steel containing 0.7% chromium 0.7% manganese, 1.8% nickel and 0.25% molybdenum
Ejector box	cast iron
Front platen	medium carbon steel
Metal injection sleeve	chromium–molybdenum steel
Plunger	0.15% carbon 0.7% chromium steel
Plunger rod	0.15% carbon 0.7% chromium steel

The list shows a skilled engineer's appreciation of the available materials and it is interesting that even today the metals used in die casting machines are similar to those selected by Doehler. For example, *Table 21.1* shows the materials used in a popular medium sized hot chamber machine as constructed in 1938 and a similar model constructed in 1981, this having been illustrated in *Figure 11.2* on page 99.

TABLE 21.1

Part description	1938 EMB 12	1981 EMB 12 B/C
Base	Cast iron	Mild steel fabrication
Metal container	Chromium cast iron	Meehanite
Fixed platen	Mild steel	0.4% carbon steel
Moving platen	Semi–steel	Cast steel 0.25% C
Tie bars	0.4% carbon steel	EN 24 nickel chromium molybdenum steel
Toggle links	Semi steel	Cast steel 0.25% C
Toggle bracket	Semi steel	Cast steel 0.25% C
Toggle pins	Case–hardened steel	0.5% carbon steel hardened
Bush and plungers	Nitralloy	EN 41a steel, nitrided

It is clear that most of the components of a die casting machine are amply strong when manufactured of carbon steels, low alloy steels, or alloy cast iron. The tie bars take some stress but low alloy steels are usually adequate. For example in the Castmaster 600 ton cold chamber machine *(Figure 21.1)* tie bars are of a steel with 0.4% carbon and from

Figure 21.1 Castmaster 600 tonne cold chamber machine. (Courtesy Markham and Co. Ltd.)

0.6 to 1.0% manganese, but if the operating conditions of such a machine indicate that the tie bars will be highly stressed, an EN 24 steel is used, containing carbon 0.36–0.44%, silicon 0.1–0.35%, manganese 0.45–0.7%, nickel 1.3–1.7%, chromium 1.0–1.4% and molybdenum 0.2–0.35%. Similarly to the hot chamber machine the links on the 600 ton Castmaster are of cast steel, the composition being carbon 0.15%, manganese about 1.3%. Even in a large machine such as the 2200 tonne Italpresse illustrated in *Figure 21.2,* the tie bars are of a steel with 0.35% carbon, 0.9% manganese, 0.4% silicon and 1.1% chromium, surprisingly close to the material selected by Doehler. Platens and toggle links are of a 0.3% carbon steel.

Figure 21.2 Italpresse 2200 tonne cold chamber machine. (Courtesy Fry's Diecastings Ltd.)

Shot sleeves and plungers for cold chamber machines

Molten metal ladled into the shot sleeve is injected into the die cavity using a water cooled plunger which transmits the intensification pressure to the solidifying casting. The plunger undergoes similar thermal and mechanical stresses and attack from the molten metal; suitable materials must be selected for shot sleeves and plungers to give maximum life and minimum maintenance. Generally the shot sleeve is manufactured from a steel which is nitrided to a depth of about 0.25mm with surface hardness up to 70 HRC.

Plunger tips require high mechanical strength and toughness with good resistance to fatigue, shock, wear and creep. Several materials are used, including Meehanite, nitriding steels, and a variety of copper alloys. Meehanite is relatively cheap and more resistant than nitriding steels to attack from molten aluminium, but suffers from lower toughness and fatigue resistance. Some improvement in properties has been obtained by spraying the tip with bronze weld and then grinding it to fit in the sleeve, but poor adhesion of the weld material can limit tip life.

Nitrided steels are sometimes used for plunger tips but they have a tendency to embrittle and to develop surface cracks under the conditions of heat and abrasive wear. Being of the same material as the sleeve with a hardness of 70 HRC, additional lubrication is required to prevent seizure and galling. These problems can be reduced if the hardness of the two components are significantly different, and this has led to the use of copper alloys for plunger tips with hardness levels up to about 40 HRC.

Allowances must be made to compensate for the higher thermal expansion of copper alloys compared to ferrous materials; larger clearances are provided between the outside diameter of the plunger tip and the bore of the hardened and ground sleeve. Tolerances of about 0.01 mm/cm are usually adequate and, under the operating pressures used, the high ductility of copper alloys provides a self-peening action on the surfaces of the tip, thereby maintaining a close fit within the sleeve. The higher thermal conductivity of copper alloys compared to nitriding steels allows improved tip cooling and heat dissipation which reduces distortion under repeated heating and cooling cycles. The relatively easy machining allows the tips, when worn, to be machined to the next lower size.

Aluminium bronzes and nickel–aluminium bronzes, having thermal conductivities very similar to nitrided steels, were used for a time for plunger tips but they were not completely satisfactory as bearing materials. Subsequently various beryllium bronzes were used; they had the advantage of higher thermal conductivity than aluminium bronze and they can be heat–treated to provide increased strength and hardness.

Beryllium–copper alloys known by their trade names of Ampcolloy 83–20 containing 2% beryllium, and 91–20 containing 0.5% beryllium, 1% cobalt and 1% nickel, have been used successfully in many parts of the world. Both of these alloys are precipitation hardened by low temperature heat-treatment. Their properties are shown in the *Table* below. Beryllium is an expensive element and all beryllium compounds are toxic. Recommendations are given for protection against injurious dust and fumes which may arise in the manufacture of copper–beryllium alloys and in the fabrication of components made from them. These problems led to the development of Ampco Metal in the late 1970s of a beryllium free plunger tip material known as Ampcoloy 940. It is a copper–silicon–nickel–chromium alloy which offers high strength and toughness and consequently is being used successfully by many die casting companies.

A modified low beryllium alloy known as BB1–MOD, developed by Metallindustria of Milan has proved to be similar in performance to the competing alloys. It contains 0.6% beryllium, 1.5% nickel and 0.8% cobalt. This slight reduction of the expensive cobalt leads to a cost saving. The following, *Table 21.2*, shows some of the materials used for plunger tips, with relevant properties.

TABLE 21.2

Plunger tip material	Thermal Conductivity (CGS units)	Coefficient of thermal expansion ($^{\circ}C \times 10\text{-}6$)	Tensile strength at room temperature (N/mm^2)	Hardness	
				Brinell	Rockwell C
Nitrided steel	0.12	11.5	1000	780	70
2% beryllium–copper heat treated (83–20)	0.30	17.0	1170	370	40
Copper with 0.5% Be 1% Co 1% Ni heat treated (91–20)	0.45	17.8	650	210	17
Copper–silicon–nickel chromium alloy (940)	0.45	17.5	630	180	16
Copper with 0.6% Be, 1.5% Ni 0.8% Co. heat treated (BBl-MOD)	0.50	17.8	760	230	20

Hydraulic fluids

Almost without exception, die casting machines are now being offered with hydraulic fluids. Indeed, many older machines originally designed for mineral oils have been converted to accept the less flammable water glycols and phosphate esters. Because of the poor lubricating qualities of fire resistant fluids compared with mineral oils, a much higher load is put on machine pumps, where even the smallest contaminating particles can lead to damage of the hydraulic components. Modifications to die casting machines for use with these fluids include reducing the speed of the pump drives, enlarging the cross sections of the suction pipes and improving filtration. With a corresponding high standard of maintenance of the fluids in use, hydraulic components can achieve a service life comparable to that using mineral oils.

Several materials and finishes, for example plating and painting, which are suitable for mineral oil systems, may not be compatible with fire-resistant fluids. The following table is given as a general guide, the term 'compatible' indicating no reaction between the fluid and the material. Common industrial paints often used in machines employing mineral oils, are affected by all types of fire resistant fluids. The paint must be removed from all internal hydraulic surfaces before the new fluids are used. Similarly the large number of different sizes of seals within a machine's hydraulic system that were compatible with mineral oil are normally suitable for water glycol but must be changed for phosphate ester fluids. Viton (a fluorinated polymer) is the recommended seal material when operating with phosphate ester and is shown in *Table 21.3* to be compatible for all types of fluids. Difficulties can arise when a group of machines operate with different

TABLE 21.3 Fire resistant hydraulic fluids; compatibility with system materials

	Water Glycol	Phosphate Ester
Ferrous metals	C	C
Copper and brass	C	C
Bronze	LC	C
Zinc	NC	C
Lead	NC	C
Cadmium	NC	C
Aluminium	LC	C
Magnesium	NC	C
Nickel	C	C
Common industrial oil-resistant paints	NC	NC
Epoxy and phenolic paints	C	C
Vitreous enamel	C	C
Acrylic plastic	C	NC
Phenolic plastic	C	C
PVC	C	NC
Nylon	C	C
PTFE	C	C
Natural rubber	C	NC
Neoprene rubber	C	NC
Butyl rubber	C	C
Viton	C	C
Leather	NC	LC
Cork, including rubber-impregnated	NC	LC

C = compatible; NC = not compatible; LC = limited compatibility fluids and dies requiring the use of hydraulic core-pulling cylinders are transferred to machines operating with a different fluid. Most types of sealing material in mineral oil systems tend to swell when used with phosphate ester; therefore all seals in core-pulling cylinders should be examined regularly and, if necessary, changed to Viton, after draining the cylinders. Although Viton is more expensive than other seal materials, it is often specified when both water glycol and phosaphate ester fluids are in use. Such standardization reduces the number of stock items and avoids any error of using the wrong materials.

Overheating and contamination are the main causes of hydraulic fluid degradation. Temperature control, using heaters and water-cooled heat exchangers is discussed on page 206. The hydraulic system designers will advise on the best materials available to achieve a high standard of filtration, using appropriate filter elements compatible with the fluid used.

Chapter 22

Instrumentation

In the production of pressure die castings, molten metal is held temporarily in a container; a plunger forces the metal into a steel die, slowly at first but with increasing speed as the injection proceeds; then the casting is rapidly chilled. Thus within the space of a few seconds the metal passes from the molten state, through a condition of very rapid movement, until it is suddenly cooled and becomes solid. The behaviour of the metal cannot be seen, but the forces exerted on it and its sequence of changes during the process, can be measured.

The days when the control of the process and the quality of the casting were in the often expert hands of the die casting operator are becoming part of history; the metallurgist, technician and electronics engineer are now working together to base the industry on scientific principles[1]. This transition from art to science became necessary to obtain higher quality levels, and to compete with other processes. The need for instru-

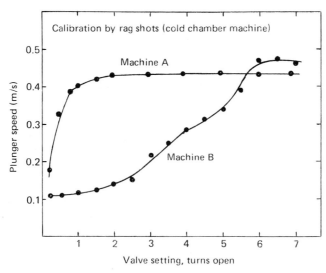

Figure 22.1 Effect of plunger speed valve setting on two machines indentical in both appearance and manufacturers specification. (Courtesy BNF Metals Technology Centre)

mentation was emphasized because of the difficulty of attracting skilled operators to work in hot and arduous conditions.

The first stage of questioning may be illustrated by a simple test[2] carried out by the then British Non-Ferrous Metals Research Association (Now BNF Metals Technology Centre). The valve settings on two apparently identical cold chamber machines were checked against the plunger speeds. Tests were made under dry conditions without metal, using a small quantity of rag in the injection sleeve to cushion impact at the end of plunger travel. This allowed the full speed range to be studied without the added problem of metal splashing from the die or difficult ejection of incomplete castings. Comparing the different results shown in *Figure 22.1,* machine A valve gave virtually no control. Machine B valve, whilst giving a better response, was seen to have all its control function between the second and sixth turns.

The first users of instruments were exploring the many process parameters, some of which are interdependent. A greater understanding of the relationship between them led to control systems to optimize production rates and quality levels under a given set of conditions. The development of instrumentation during the past 25 years has passed through three stages. First the existing methods of control were tested and experiments conducted to discover whether conditions could be reproduced. Then in the 1960s and early 1970s experiments were carried out in the use of instrumentation, to compare methods, to determine which parameters were important and which were not so vital. The past 10 years have seen great advances in the efficiency and compactness of instrumentation so most new machines are provided with inbuilt controls. Today the application of this knowledge has advanced to the stage where minicomputers are incorporated in the die casting machine to record operating conditions. These systems can also be fed to a central computer, both for operation and statistical purposes, allowing centralized control of a number of machines. These programmable controllers on die casting machines offer reliability and flexibility of sequencing which are particularly important where ancillary equipment such as ladles, casting extractors and spraying systems operate automatically within the machine cycle.

Since logic controllers have diagnostic capabilties, present efforts are being directed to the continuous measurement of operating parameters of the machine[3]. Microprocessor-based systems are capable of accepting incoming information and, after comparing it with stored data of optimum conditions, an output signal can be given either to control the machine parameter or give an alarm signal if the incoming information is not within selected limits.

During the initial die design stage, technology is now available to improve productivity. Computerized systems[4] offer a new approach to die design in which the route taken by the molten metal through the die is simplified to the point of being determined mathematically. By matching the capacity of the machine with the die characteristics, it can be determined quickly whether the chosen conditions will produce sound castings with good surface finish. This has led to reducing the amount of surplus metal associated with runners and overflow systems. With the facility of a thermal programme, a designer can also estimate where internal die cooling is required and calculate the cooling effect of any cores.

The Light Metal Founders' Association, which represents the majority of the aluminium castings industry in the UK has launched several projects in research and technology transfer[5]. With support from the Engineering Materials Requirements Board of the

Department of Industry the work covering aluminium pressure die casting is making use of existing forms of instrumentation with a view to developing inexpensive, simple to use, reliable and robust equipment to monitor melt quality and to improve productivity and product reliability. The initial emphasis on melt control is developing equipment for the measurement of hydrogen gas levels. Prototype units have shown that the level of dissolved hydrogen in aluminium alloy melts can be determined accurately. The test procedures are simple but conditions are controlled precisely. Hydrogen content can be measured through a probe immersed in the melt or from a melt sample and the gas content is shown as a digital display. The equipment should not only be useful for routine quality control but will also enable measurement of hydrogen pick up or loss during various stages of metal processing.

Research combined with technology transfer forms the basis to improve product quality and consistency, and raise casting output through the development of improved procedures for the selection and control of conditions for injection of the molten aluminium into the die cavity. A study is being made of the parameters involved in the injection process on die casting machines and the knowledge gained will be used to produce a set of guidelines for the industry to use in selecting and controlling injection conditions on new and existing dies.

This chapter now examines several important process parameters – the type of instrumentation used to define them and how the findings have been brought together in the development of complete control systems. The following variables need to be measured:

1. Metal control
2. Machine parameters – locking force
                                 plunger velocity
                                 plunger displacement
                                 plunger pressure
                                 hydraulic fluid temperature
                                 machine cycle time
3. Die parameters –   die temperature
                                 die coolant flow
                                 internal die pressure

Metal control

Since die casting is a thermal cycling process, temperature control of the liquid metal is an important factor. This determines the ability of the die to fill, the amount of sticking or soldering of the casting to the die, and the surface finish of the casting. High metal temperatures undoubtedly give greater fluidity, which allows easier filling of thin sections, but it can lead to metal oxidation, loss of lower melting point constituents and is wasteful on power. Lowering metal temperatures to a level insufficient to cause any marked deterioration in surface finish has been shown to improve pressure tightness of aluminium[6] and zinc die castings[7]. Metal injected at low temperatures may permit the casting to be partially solidified while the gate is still molten so that injection pressure can assist feeding of the casting.

The effects of temperature on the metal, the machine, the die and other equipment are different for each alloy. For example, too high a metal temperature in zinc die casting leads to galvanizing of the die and loss of magnesium in the alloy. With aluminium, too high a temperature leads to oxidation and to deterioration of the crucible. On the other hand, too low a temperature encourages compounds of silicon, iron and manganese to settle out in the melt as a sludge. If transferred into the casting, these metallic segregates form hard spot inclusions which can cause machining difficulties (*see* page 73). With magnesium, too high a metal temperature causes oxidation and, if taken too far, to metal burning. In hot chamber production of magnesium die castings it leads to deterioration of piston rings. In brass die casting, too high a temperature leads to reduced die life, zinc fume and furnace deterioration.

Locking force

As a measure of a machine's potential, locking force determines the ability to hold two halves of a die together with sufficient force to contain the pressures exerted from injected metal, which can exceed 200 MPa (30 000 lbs/in^2). When locking force is too low, flashing from molten metal can occur, while excessively high locking forces can lead to premature failure on tie bars and unnecessary wear on other equipment, including the die. Calibrations can be made as a simple guide to aid machine setting. *Figure 22.2* shows the relationship recorded between locking force and hydraulic pressure in a cold chamber locking cylinder.

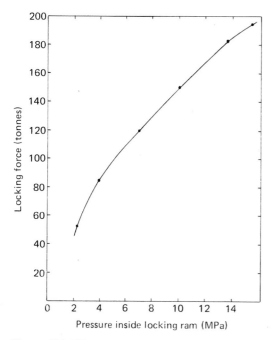

Figure 22.2 Effect of line pressure on the locking force of a 200 tonne cold chamber machine. (Courtesy of BNF Metals Technology Centre)

Plunger velocity and displacement

Displacement of the injection plunger usually occurs in two stages. On the first slow approach stage metal is brought up to the die ingate. In the second stage, molten metal is pushed into the die cavity in a short filling time with high injection speed. If the slow approach is too long, liquid metal reaching the ingate at low velocity can partly solidify, causing reject castings, whilst if the slow approach is too short, air is entrapped during the die filling. The cavity filling time in the second stage must be equal to or shorter than the time taken for the metal to solidify in the die cavity which is dependent mainly on the section thickness of the casting, metal and die temperature. For a given gooseneck or shot sleeve configuration, die cavity filling times are determined by plunger velocity during that part of the injection stroke when the metal is passing through the die ingate.

The importance of cavity filling rate on the soundness of a brass die casting is illustrated in *Figure 22.3* taken from the report by BNF Metals Technology Centre[8] on a brass mixing valve. In general, the main effect of lower filling rate is to reduce metal turbulence which minimizes the entrapment of gases. In initial tests with filling rates as high as 550–850 cm^3/second, it was found that a minimum slug length of 5.5 cm was required to achieve acceptable porosity below 1%. Then, after reducing the filling rate to 270 cm^3/second, the slug length could be reduced to 3.6 cm, while still achieving the acceptable porosity level. Any reduction in the volume of metal required for each shot can show significant savings in remelt costs. As might be expected design factors such as gating and venting interact with the filling rate effect.

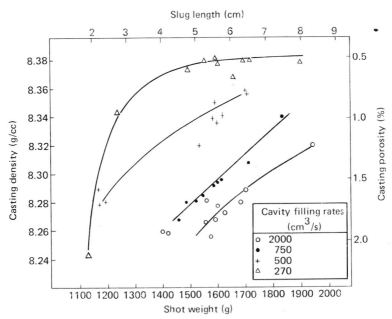

Figure 22.3 Effect of large shot weight variations on the density of a brass pressure die casting using four different injection speeds. (Courtesy BNF Metals Technology Centre)

Plunger pressure

The soundness of a die casting is improved by increasing the pressure transmitted during solidification, which will reduce the porosity due to metal shrinkage and entrapped gases. This can be done either by applying more pressure to the injection plunger or by increasing the slug length so that the metal will stay molten longer and assist pressure transmission from the plunger to the casting. A significant reduction in porosity in a small aluminium cylinder head casting is illustrated in *Figure 22.4*. Increasing the final pressure in the injection cylinder from 6.9 MPa to 13.8 MPa reduced casting porosity by almost 40%. The applied pressure at the end of cavity fill and during solidification should be as high as possible within the locking force limits of the machine. Reducing the size of the injection plunger may also be necessary to achieve the desired injection pressure on solidifying metal.

Although there is no evidence to indicate that excessive pressures are harmful to the solidification process, pressure surges transmitted through the hydraulics of the machine can affect casting quality, while excessively high pressures can cause die flashing and lead to damage of hydraulic injection systems. Pressure variations at the rod end of a shot cylinder affect the performance rating of the machine hydraulics and determine whether a satisfactory pressure level can be achieved at the commencement of each cycle.

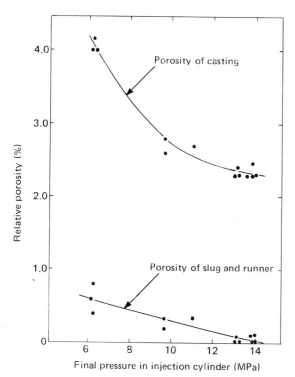

Figure 22.4 Effect of hydraulic injection cylinder pressure during solidification on the porosity of a small aluminium alloy cylinder head casting. (Courtesy BNF Metals Technology Centre)

A reduction in casting reject rate from 12% to 3% was obtained[9] after taking a machine injection trace during the production of an aluminium alloy casting showing incomplete cavity fill and porosity in a thick-walled area. Delayed plunger acceleration at the beginning of cavity fill and poor pressure build up at the end of injection was found to be due to a wrongly fitted fluid restrictor in the machine. Correcting the fault resulted in improved casting quality and the savings made in a two month period from reduced rejects paid for the diagnostic instrumentation.

It was reported by D. Askeland and D.S. Patel[7] that on hot chamber production of a zinc alloy test plate with dimensions 12.7 x 7.6 x 0.3 mm, increasing injection pressures up to 50% made a substantial reduction in gas and shrinkage porosity, and the importance of maintaining the pressure during solidification was also demonstrated. When the pressure was not maintained long enough a vacuum was created in the injection chamber due to the return of the plunger. Because of this vacuum and the pressure exerted by the compressed gases in the casting, liquid metal was forced from the casting back through the ingate, leading to concentration of shrinkage and porosity near the ingate, and poor soundness throughout the remainder of the casting.

Hydraulic fluid temperature

During the commencement of die casting machine operation, the temperature of the hydraulic fluid increases, with a corresponding decrease in viscosity. Until the fluid reaches its optimum temperature, its changing viscosity can affect the consistency of plunger speeds. Excessive overheating can lead to breakdown of the fluid components with a resultant, high and sometimes catastrophic wear of pump parts. Proper maintenance of the fluid is important in this respect and heat exchangers are now commonplace in machine hydraulic circuits. On larger machines which need several hours to warm up, thermostatically controlled heaters are sometimes used, so ensuring that hydraulic fluid is at the optimum viscosity before casting production commences.

Hydraulic fluids

The need for safety has led to the development of fire resistant fluids which are less flammable than the mineral oils which they replace. Nonflammable fluids can be grouped into water and non-water-containing types. The water-containing emulsions of mineral oil have been going out of favour during the past decade and polyglycol solutions containing about 40% water are now widely used. Of the non-water-containing fluids, those based on phosphate ester are homogeneous fluids of liquid organic compounds. Phosphate ester is very sensitive to water contamination where the presence of more than 0.1% causes the formation of acidic compounds which render the fluid useless.

Depending on the operating temperature of water glycol fluids, water will be lost through evaporation, increasing the fluid viscosity; any additives, particularly alkaline corrosion inhibitors, will be partly lost in the evaporating water. The amount of water loss should be checked periodically by measuring fluid viscosity and making additions of distilled water according to the manufacturer's recommendations. Checking pH values for alkaline content will indicate if there is a need to replace lost additives, but if a recommended maximum fluid operating temperature of 40°C is not exceeded, losses will be minimized.

Phosphate ester fluids have higher viscosity properties than water glycols. The ideal operating temperature is 50°C, with 90°C being the highest permissible level. Above this degradation occurs, rendering the hydraulic fluid unserviceable.

Machine cycle time

Affected by the design of the die, a machine casting cycle time is also dependent on other variables. Production rates are often monitored as a measure of improved casting performance that may be achieved for a certain standard of casting quality. For a given set of machine and die conditions, optimization of the cycle time can be achieved by the use of a die temperature-cycle time controller as discussed on page 215.

Die temperature and die coolant

In order to maintain optimum production rates, die temperature is important in terms of heat transfer. Temperature distribution is generally affected by regulating the flow of cooling media through the die cooling channels, although this area is one of the least precisely controlled variables in the die casting process. The main die areas to control are those of the cavity face at injection and ejection. Face temperatures at injection have a marked effect on surface finish of the casting whilst face temperatures at casting ejection affect dimensional tolerances of the castings, and the formation of surface blisters.

Internal die pressure

The measure of internal die pressure will indicate if molten metal is reaching all areas of the die cavity before solidification begins. This is an important variable to monitor in the production of thin-wall castings, where the tendency for premature solidification is greater than in heavier section castings.

Full intensification pressure provided by a machine's injection system while a casting is solidifying is often measured with dial gauges; then by reference to plunger size and die thermal patterns, the pressure within a die cavity can be calculated. Alternatively a direct reading of cavity pressure can be made with a transducer incorporating strain gauges mounted behind an ejector pin. Should maximum cavity pressure not be achieved it could indicate that areas of the die, or metal are too cold, or the injection plunger is sticking in the shot sleeve, or the machine's injection system is not operating correctly.

Methods of measurement

Having identified important variables in the die casting process, the next stage is to use instruments to measure these variables in the areas listed below.

1. Metal control and furnace conditions.
2. Machine setting when a die is put into service.
3. Monitoring parameters during production.
4. Establishing the causes of casting rejects or poor machine performance.

Although standard ranges of instruments have long been used to measure varying conditions by employing dial gauges, temperature recorders and similar equipment, more elaborate instrumentation is required which will be sensitive over the rapid cycle time of pressure die casting. The introduction of high speed recording equipment was sometimes met with reluctance by foundrymen, due mainly to the unfamiliar appearance and terminology used in its application. Over recent years, such instruments have been generally accepted as it was realized that they are easy to use, although an efficient electrical department is needed to maintain, repair and calibrate them.

Furnace conditions

Most modern furnaces are temperature controlled, using standard pyrometry and controllers to adjust automatically the heat input. With recent advances made in microprocessors, monitoring of combustion conditions in fossil-fuel furnaces has become a practical possibility; *Figure 22.5* shows a hand-held monitor costing about £500. The extension probe, which contains a thermocouple, is inserted into the furnace flue. Gas temperature is measured and a pump draws a sample of gas to an electrochemical oxygen sensor. The microprocessor uses this information and, taking into account the type of

Figure 22.5 Hand held fuel efficiency monitor. (Courtesy Neotronics Ltd.)

Figure 22.6 Trolley mounted furnace efficiency monitoring system. (Courtesy BNF Metals Technology Centre)

fuel used, calculates the percentage combustion efficiency of the furnace which can be retained in a memory and displayed with readings of temperature and oxygen content.

A trolley mounted system, as illustrated in *Figure 22.6* and costing less than £10 000, has been designed to make regular readings on a number of furnaces without the need to instal permanent instrumentation on each furnace. Direct indications of the fuel flow rate, heat loss through the flue and combustion conditions are shown on a digital display, together with an indication of the remedial action necessary to improve burner combustion. Where a permanently installed fully automatic control system is required, for example on a central bulk melting furnace, the computer in the trolley can be incorporated within the furnace control system. In addition to carrying out all forms of data monitoring plus necessary calculations, the unit will function as a control system so that fuel and air to each burner can be adjusted automatically to obtain optimum efficiency for a chosen melt rate.

Metal analysis

Many companies which purchase alloys in ingot form against standard specifications either have their own in-house analysis facilities or use service laboratories to check their alloy compositions. A technician and an assistant carrying out wet chemical analyses of the major constituents with an atomic absorbtion spectrometer for trace elements (costing about £10 000) can perform up to 10 full analyses per day. One laboratory technician using a direct reading spectrometer (costing about £40 000) can handle up to 15 or 20 full analyses per hour, spot sampling, etc. can be carried out as well as analysis of incoming material. *Figure 22.7* shows a typical instrument providing a print-out of each analysis.

Figure 22.7 Direct reading spectograph for metallurgical analysis. (Courtesy Applied Research Laboratories Ltd.)

Machine conditions

It is in the die casting machine operation where managements must determine which of the many variables require measurement and how the measurements are to be made. Today, packages of instruments to monitor machine performance are assembled in a mobile trolley with quick connections and can be taken to each machine and used with minimal interference to the casting cycle. Alternatively some manufacturers are incorporating recording equipment into existing machine control cabinets. A typical mobile diagnostic trolley, costing about £4000 and incorporating an oscillograph recorder which gives a permanent record of the conditions being measured, is shown in *Figure 22.8*. Due to the rapid response required and the need to utilize the full width of the recording paper for simultaneous traces that may overlap, recordings are made on light-sensitive paper with a beam of ultraviolet light. Parameters that can be measured include:

1. Pressure exerted on injection plunger
2. Speed of injection plunger travel
3. Machine locking force.

This easy to operate instrument package consists of transducers to measure both plunger displacement and pressures, and strain gauges to check maching locking force. The measured variables are displayed, using an ultraviolet recorder with a continuously moving paper chart to provide a record for study and comparison.

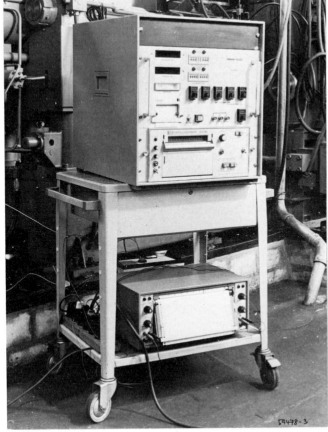

Figure 22.8 The diagnostic trolley. (Courtesy S.E. Laboratories and Buhler Bros.)

Plunger pressure and displacement

For pressure measurements, hydraulic pressure transducers are attached into valve or cylinder gauge ports in the die casting machine. The transducer shown in *Figure 22.9* consists of a diaphragm in which are bonded strain gauges that provide a voltage output proportional to the pressure applied to the diaphragm. Other pressure transducers of semiconductor, capacitance and inductance types are available from a number of manufacturers.

Many devices are available for measuring the injection plunger speed and distance of travel, while some can be obtained with a tachogenerator to record velocity directly. Photocells and similar devices have been used and, in conjunction with a detector strip attached to the plunger rod, the time required for either beam interruption or strip travel between two set points is measured. A system based on an ultrasonic transducer uses a permanent ring magnet over the transducer rod. Displacement of the injection plunger moves the magnet, which generates a continuous velocity and displacement profile signal.

Figure 22.9 Hydraulic pressure transducer shown inserted in a convenient bleed-plug socket in the injection cylinder. (Courtesy BNF Metals Technology Centre)

Figure 22.10 Wire and drum type plunger displacement transducer mounted on a cold chamber injection cylinder, with the extension wire attached to the plunger. (Courtesy BNF Metals Technology Centre)

The most popular method uses a wire and drum type of displacement transducer consisting of a flexible wire connected to a helical potentiometer as shown in *Figure 22.10*. Plunger rod movements extend the connecting wire and the potentiometer provides a voltage signal proportional to plunger rod position.

A typical injection trace using oscillographic recording equipment is illustrated in *Figure 22.11*. Plunger displacement and pressure recorded against injection time are shown for an aluminium die casting produced on a 400 tonne cold chamber machine. From the cycle start position A, the injection plunger is seen to be displaced in two

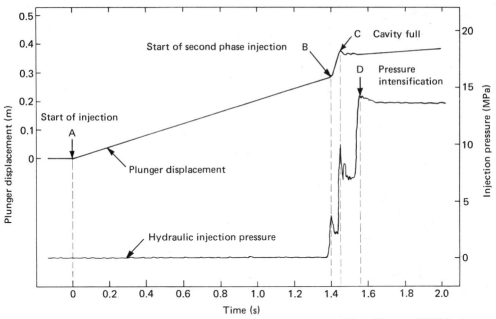

Figure 22.11 Injection trace taken on a 400 tonne cold chamber machine. (Courtesy BNF Metals Technology Centre.)

separate stages corresponding to the first and second—stage injection phase. The first stage terminating at position B after a time of 1.4 seconds represents a slow prefill of metal up to the die gate to reduce turbulence and allow time for venting of air from the die cavity. This first stage plunger speed determined by the gradient of the trace was calculated at 0.2 m/second. The second injection stage occurred between trace positions B and C where a faster plunger speed of 1.4 m/second was calculated as the die cavity filled with metal. Beyond this cavity fill position, the small amount of slow plunger displacement is governed mainly by the total shot weight and represents the final compaction of metal in the die cavity as solidification advances.

The corresponding hydraulic pressure trace in the injection cylinder shows a small rise in pressure at point B where the plunger speeds are changed, and indicates the various restrictions that are met as metal passes through the die ingates and runners. Pressure builds up rapidly at position C where the plunger reaches the end of its stroke and fluctuating pressure levels indicate a water hammer effect. A third injection phase leading to high pressure build up to assist in consolidating the cast metal occurs after a short delay of about 0.1 second after the casting was filled, shown at position D.

An injection system called 'Parashot', designed by the Swiss manufacturer Buhler and described by P. Koch[10], is claimed to reduce turbulence of molten metal in the injection sleeve as the plunger advances. Hydraulic servosetting valves give a constantly accelerating injection movement which, compared with a conventional constant speed movement, will reduce the generation of waves on the surface of the molten metal in the injection sleeve as the plunger advances. Hydraulic servo setting valves give a constantly casting density, a more homogeneous structure and a better surface finish. Comparison of plunger displacement traces, shown schematically in *Figure 22.12,* illustrates the continued acceleration represented by the parabola form on the graph.

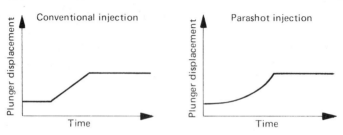

Figure 22.12 Comparison of conventional and Parashot injection system

Traces produced with diagnostic equipment when castings of acceptable quality are being made provide standards which can be referred to at some future date. This is a valuable aid for checking machine parameters and ensuring that satisfactory injection conditions can be reproduced. Faults on the machine and die can be localized precisely so that they can be eliminated. The valve calibration illustrated in *Figure 22.1* was obtained using diagnostic equipment of the type illustrated.

Machine locking force

Locking force can be measured by the hydraulic pressure in the locking cylinder when the die is closed, or by measuring the strain on each of the tie bars, using strain gauges and extensometers and calculating the amount of stress. Alternatively locking force can be checked using a 'dummy die' in the form of a load cell incorporating strain gauges. *Figure 22.13* illustrates a typical load cell in position on a die casting machine. Strain gauges attached to the load cell are connected to dial gauges on an oscillograph recorder. Locking force is plotted at various settings of the tie bar nuts until maximum lock is recorded. Any deviation from a linear plot as the tie bar locking nuts are varied in position indicates a defect in the machine locking system.

Figure 22.13 Load cell for lock testing of die casting machines. (Courtesy BNF Metals Technology Centre)

Hydraulic pressure measurement is a more direct means of assessing locking force. If supplemented by the die load cell or tie bar strain measurements, it is possible to detect unequal load distribution on the tie bars or toggles due either to the die faces not being parallel or the die itself being positioned off centre. Continuous measuring during production will indicate wear on linkage pins or bushes which cause uneven straining to both the die and the machine locking system.

Die temperature

Die temperature can be controlled either by regulating the flow of cooling media through the die, or by using a die temperature/cycle time controller, allowing the temperature of the casting instead of the machine timing to control the die cooling cycle. The basic instrumentation required includes a thermocouple, a temperature controller and a solenoid-operated fluid valve. Response from the thermocouple positioned near to the die face cavity is monitored by the temperature controller which, in turn, operates the opening or closing of a solenoid valve to allow a controlled flow of cooling fluid, normally water, into the die. This system of control cannot be used on all waterway lines, particularly those close to the surface of the die, since thermocouple responses in these areas would fluctuate widely and water flow may become out of phase with heat extraction.

A further development in die temperature control[11] uses heat transfer oils and nonflammable thermal fluids for both die heating and cooling. Although methods of temperature control using oils have been known for a long time, the main applications have been in plastic injection moulding and it has been used only recently in pressure die casting. The principle of the system is to heat the die to the required production temperature and to maintain this temperature automatically by using oils which are pumped through a heat exchanger at temperatures up to 300°C and circulated in the die cooling channels. This method has advantages over open flame or radiant heaters placed between the two die halves since the heating is more uniform and avoids the risk of high thermal stresses being introduced when casting production commences. Temperature control is achieved using thermocouples in the fluid return lines, which signal valves to pass the circulating fluid either through heating or cooling sections of the exchanger.

Depending on the degree of thermal control required in a particular die, several heating-cooling circuits can be selected to operate independently. Since the heat extraction rate of oils is less than that of water, additional cooling channels are normally specified, while in selected die areas requiring a greater amount of heat extraction, separate water cooling can be provided.

The basic principle of die temperature-cycle time control is to allow the temperature of the casting rather than a fixed machine cycle time control the solidification and cooling phases of the casting within the die. Opening the die when the temperature of the casting is at a preset level assists in maintaining a constant die temperature, optimizes casting rate and reduces dimensional variations of the cast part.

Initially an area within the die cavity is chosen for positioning the thermocouple which is usually associated with the last parts of the casting to solidify. Temperature response must be rapid and this involves positioning and brazing the thermocouple flush with the die cavity face. In practice it has been found acceptable to locate thermocouples in small cores or ejector pins followed by brazing with a nickel–chromium alloy to effect rigid location and good thermal contact. Mineral insulated nickel–chromium,. nickel–

aluminium thermocouples with 1mm diameter stainless steel or 'Inconel' outer sheaths perform satisfactorily. In operation the machine's own cooling timer is set at a low minimum safety level, allowing the control relay to give the die open signal, as the sensed die face temperature falls to a preset point. Additional control relays are often used where several thermocouples are required to measure other critical casting areas, including the slug, to prevent bursting when the die opens.

References

1.   BOOTH, S.E.; HILL, T.B.; STREET, A.C.
      Controlling and monitoring the diecasting operation. *The Diecasting Book.* Portcullis Press Ltd, Redhill, Surrey and American Die Casting Institute, Des Plaines, Illinois. Chapter 6 (1977)

2.   *BNF Guide to Better Aluminium Diecasting.* BNF Metals Technology Centre, Wantage, Oxford-shire.

3.   MOORE, J.
      Implementing a process control program. *Die Casting Engineer,* **21,** (4) 28 (July/August 1977)

4.   KELLOCK, B.
      Computers mould diecasting's future. *Machinery and Production Engineering,* **137,** (3541), 188 (10th December 1980)

5.   *Private communication,* Dr. J.A. Rogers, Light Metal Founders' Association

6.   GARBER, L.; DRAPER, A.B.
      The effect of process variables on the internal quality of aluminum die castings. *Society of Die Casting Engineers* Congress paper G-T79-022 (1979)

7.   ASKELAND, D.; PATEL, D.S.
      Factorial analysis of some variables and their optimization for the production of sound die castings. *Society of Die Casting Engineers* Congress paper G-T77-013 (1977)

8.   *BNF Guide to Better Brass Diecasting.* BNF Metals Technology Centre, Wantage, Oxfordshire

9.   KOCH, P.
      Die Casting Metrology. *Technical Information Brochure, No. 4.* Buhler Brothers Ltd., Uzwil, Switzerland

10.   KOCH, P.
      Parashot system for turbulence-free injection of metal. *Die Casting Engineer,* **18,** (6) 44 (November/December 1974)

11.   WIGHTMAN, D.E.
      Temperature control in the die casting process through the use of a heat transfer system. *Society of Die Casting Engineers* Congress paper G-T77-066 (1977)

Chapter 23

Lubrication developments

Die face lubricants perform a parting function between cast metal and die surface to assist metal flow and casting release, reducing the tendency of die cast metals to weld to the die. In zinc die casting, this effect is generally known as 'galvanizing', while with the other non-ferrous alloys it is called 'soldering'. Moving die parts such as slides, cores and ejector pins are lubricated to prevent seizure and wear. Plungers in cold chamber machines need lubrication to avoid seizure. Moving parts of a machine operate under high pressure and loads, so they must be protected by lubricants to avoid excessive wear. These aspects involve individual lubrication problems, so the materials and methods of application must be tailored accordingly.

During the past 20 years many changes have taken place in the compositions of die face lubricants and in the methods of applying them, stimulated by the transformation of die casting from art to science and by the development of automation. Lubricant coatings deposited on a die face need to conform to the requirements listed below.

1. High release factor: castings should part from the die easily without distortion.
2. Prevention of soldering: lubricants must form a barrier between injected metal and die.
3. Surface finish control: lubricants should not leave residues on the die face which corrode the die steel or impair the casting surface and subsequent finishing operations.
4. Casting soundness: lubricants must not develop excessive amounts of gas which can lead to porosity and reduce the pressure tightness of castings.
5. Health and safety: ingredients which are dangerous to health or cause unsafe or unpleasant working conditions must be avoided.

Some of the earliest die face lubrication practices will be examined before illustrating how modern techniques have developed from better understanding of the die casting process and the basic requirements of a lubricant.

Early die face lubricants

When die casting was in its infancy, it was customary to acquire any form of grease, oil or wax and swab the die areas, using a brush, glove or oil rag. Anything to hand in a tin, box or carton which resembled a substance with lubricating qualities was used, even reclaimed engine oil. Many operators possessed their own favourite compositions and soon realized that graphite in fine powder or flake form, when added in small amounts to their secret formulations, improved casting release. It was an efficient parting agent at high temperatures and its use represented one of the first major improvements. Graphite, although possessing good release properties, caused staining and discolouration of the castings but this could be reduced by the addition of aluminium powders to the lubricant.

Oils, greases and additives applied by hand swabbing did not cover all the important areas of the die adequately, although expert die casters were capable of producing remarkably clean castings. Compressed air was found to be a better way of distributing lubricants and at first a lubricant-coated rag was attached to the open end of an airline held by the machine operator. Without any directional control, overspraying occurred and both the operator and the surrounding environment became dirty. Paste mixture remained the standard die face lubricant for many years and still finds occasional use.

The development of paste-type lubricants

All modern sprayable lubricants, with or without pigment, are dilutions of pastes in solvents. The constituents include fluid components, usually a mineral oil, which acts as a carrier for various additives and additional pigments which are the heart of this type of lubricant.

Fluid components

A brightstock oil which has been solvent refined to remove any health hazard associated with its aromatic content is generally used. Brightstock oils, often referred to as 'straight oils', provide hydrodynamic lubrication for automobiles but, in die casting, they offer hydrostatic lubrication properties, often referred to as release, parting ability or lubricity. By acting as carriers for the lubricants' additives and pigments, their ability to wet and spread over the die face assists in dispersing the lubricant coating.

Additives

The main function of extreme pressure (EP) additives in a lubricant is to prevent metal-to-metal welding under sliding conditions and high loading pressures. These lubricants are used in gear oils and metal drawing machinery in conditions of sliding friction. Conventional EP additives have been included in die casting lubricants but, as the process does not have many analogous sliding friction situations, their merit is limited and their true function uncertain. Most EP additives undergo a chemical reaction with the hot die surface. Organic sulphides, chlorides and phosphides are of this type and the thin lubricating film produced by the chemical reaction becomes firmly bonded to the die steel. Although possessing high load-carrying properties, this film will gradually wear away under sliding conditions, only to be replaced by further reaction. Due to this chemically agressive nature and product build up, their use in die casting has gradually declined. Acidic pastes are available which react in a similar way and can be applied selectively to a die by hand swabbing but their use must be monitored carefully.

Other extreme pressure additives improve sliding characteristics under load in a physical manner without becoming chemically bonded to the die face. Graphite and molybdenum disulphide fall into this category; the latter, having lower temperature stability and producing sulphur dioxide on decomposition, does not normally form part of a die lubricant except as a bonded coating referred to on page 230. The lamellar structure of graphite allows the microscopic particles to align on the die steel surface forming a smooth refractory barrier film between the molten metal and die steel. This is necessary, for example where a casting, following solidification shrinkage, is held tightly on deeply recessed cores having a limited amount of taper.

Pigments

The third category of materials used in die lubricants, generally classed as pigments, are included to improve the separating properties of the lubricant and keep the liquid metal away from the die steel. Under the pressure and high temperature conditions within the die, a chemical reaction can take place between the cast metal and the die face where zinc and aluminium form zincates and aluminates. Under these conditions the casting tends to become welded to the die, making ejection difficult. To prevent this, it is necessary to place an inert film between the die and the cast metal. The properties of graphite make it suitable, but its use is limited, due to its tendency to discolour and become embedded in the casting surface. However for large and complex castings, graphite is frequently the only material that will provide adequate lubrication. Finely divided aluminium flake applied in a similar way will form a suitable barrier between the die face and the casting, but precautions need to be taken to ensure that build up is not allowed to occur.

Modern sprayable die face lubricants

The changed circumstances caused by mechanization, increasingly complex castings and higher melting point casting alloys led to a need for better dispersion of lubricating materials on the die face. Solvents such as diesel fuel and paraffin were the first materials to be mixed with the lubricant. Sprayable paste-type lubricants, containing relatively large amounts of graphite and aluminium pigments, were found to be deficient. The solvents caused health and fire risks; residual pigments would adhere and build up in many parts of the die cavity, cores and ejector slides. Lubricants began to be developed more scientifically in the early 1960s and there has been a transition from pigmented to the non-pigmented lubricant formulations that are setting the pace for today's major usage.

Carrier materials

These are used to dilute the lubricant concentrate and, generally with the assistance of compressed air, deposit the lubricant on the die face in a thin uniform coating. Carriers are divided into solvent fluids, which are miscible with the lubricant concentrate and those which use water as the main carrier, commonly called water-based lubricants and classed as emulsions. Solvent-based carriers used in the past have been paraffin and diesel oil, but white spirit gained favour because it evaporates from the die face without causing casting stains or deposit build up. White spirit is composed of about 30% paraffin, 58% naptha and 12% aromatics. It has a density of 0.774 and a flash point of 36°C but, when it is blended with heavier oils in a die face lubricant, the flash point of the mixture is over 200°C. Such solvents spread on hot die surfaces and remove heat but they create copious amounts of vapour, causing health and fire risks, and may lead to casting porosity. Undoubtedly, oil based lubricants would have been preferred if die casting had never been automated and if health and safety aspects were of no account. Because these changes did occur, water based lubricants had to be developed, persistently and sometimes against prejudice, until in the late 1970s it was certain that this type of lubricant was to replace the older compositions. They are used almost exclusively in the USA and are rapidly developing in Europe, so that by the early 1980s water will account for over 90% of carrier materials.

Almost all of the water-based lubricants are oil or wax in water emulsions. Oil is immiscible with water but will form an emulsion in which oil droplets are surrounded by a water film. As water is a polar compound, droplets remaining in association with it will be negatively charged to maintain electrical balance, without which the various materials would begin to disassociate. In a well-formulated stable lubricant emulsion, the particles of the immiscible phase will be small and the charge they carry relatively high, so that they do not separate from the water carrier. Upon contact with a hot die surface, the water evaporates first and, in so doing, heats the lubricating materials which are subsequently deposited on the die face. Water is finding wide use as a carrier fluid, especially now that its application is understood. It does not lead to health and fire problems, will not leave deposits on the die face and it cools the die. Lubricant dilution rates with water vary according to the application; in some cases up to 100 parts of water to one part lubricant, though on average the ratio is about 40:1. When water-based lubricants were first used, some difficulties were experienced because they did not wet or spread over the die very readily but, when this problem was understood and overcome, they became the normal while other lubricant carriers became the exception.

Lubricating materials

In contrast to the pigmented paste type lubricants containing graphite and aluminium, a high standard of die face lubrication and separation can be achieved using non-pigmented mineral oils which contain additives in solution or suspension. Lubricant manufacturers had to learn how to produce stable emulsions of the oils that had previously been dissolved in solvents; then improved release agents were introduced following the technology involved in emulsion chemistry. Commonly used materials include various oils and waxes which, when first applied to a hot die, polymerize into solid films which provide the casting release properties.

Silicones, due to their high temperature stability, can produce a thin but tenacious hydrostatic barrier film between cast metal and die steel which prevents galvanizing and soldering. When silicone oils were first introduced into die casting lubricants they were not compounded with other organic oils, so deposits formed on the die face. In addition, special degreasing processes were required for their removal to enable castings to be painted or plated. Under the conditions of high temperature and pressure to which a die lubricant is subjected, polymerization of the silicone molecule takes place to produce a viscous film which provides most of the parting function. It is believed that after a period of time an adhesive vitreous-like deposit can build up on the die face but the advantages of silicones outweigh their disadvantages. The new generation of blended silicone oils are finding some use in die casting lubrication; further details of their development are given on page 229.

Technology of die face lubrication

Diverse and interrelated chemical and physical reactions occur during the application of die face lubricants. Researches have been initiated in this field and in developing new lubricating materials but, because of the difficulty in isolating a single variable under die casting conditions, little work has been done related to the mechanism of lubrication. Statistically valid results are even more rare, although in a paper by A.B. Schmidt[1], some factors influencing the entire die lubrication system were presented, based on physical and chemical principles.

As a lubricant is applied to a hot die surface, there are changing physical and chemical conditions and the system seeks a new state of equilibrium. Depending on the molecular weight of constituents in a lubricant, vapour pressures will change as the carrier fluids evaporate. Sprayable lubricants containing volatile solvents with high vapour pressures lead to rapid evaporation on the hot die. Since the degree of die cooling is dependent on the latent heat of vaporization of the carrier fluid, and this property for water is about 10 times greater than for solvents, water-based lubricants have a greater potential for cooling dies. This large difference in thermal shock at the die face is going to have some effect on the stresses induced in the steel, but the degree of heat checking actually caused by repeated spraying of the die with water would be difficult to determine experimentally. The theoretical effects could be calculated by comparing the heat capacities of water, oil based lubricants and aluminium alloy and making some basic assumptions regarding spray volume, times of application and temperature gradients. By such calculations it can be shown that whichever carrier is used, the contribution of the lubricant to heat checking is less than 2% of the effect of the die cast aluminium alloy.

Given a separating agent with high surface tension and high pressure, the 'Leidenfrost' effect may develop, where droplets do not reach a hot die face but evaporate prematurely under the influence of radiant heat. This effect is particularly pronounced when water-based die lubricants are used. Consequently it is necessary to spray with

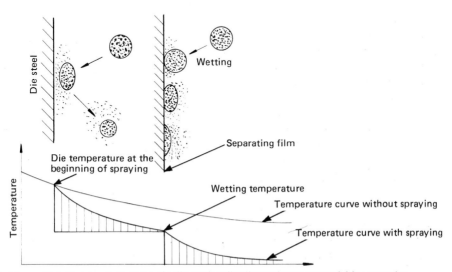

Figure 23.1 Temperature curve of a die casting die when using a water soluble separating agent

sufficient power to propel the droplets at high momentum and break through the Leiden-frost vapour cushion. Alternatively, spraying for a longer period of time until wetting begins can be employed, although it is wasteful of lubricant and time.

Figure 23.1 illustrates the effect of die temperature on the cooling properties of water based die spray. When the die temperature is high at the beginning of spraying, a vapour cushion develops, causing the lubricant droplets to bounce away from the die without wetting the surface. When the die temperature is the same or lower than the wetting temperature the droplets adhere and spread over the die face leading to rapid cooling and evaporation of the remaining solvent. Once the correct amount of lubricant to be applied has been established, dies requiring considerable cooling are sprayed for a longer period of time at high dilution, whilst for less cooling a more concentrated mixture is applied.

Lubricant material which is subsequently deposited on the die must have adequate wetting properties and not coalesce or run off the surface. Extremely thin films of lubricant are known to have good bonding qualities to a die surface and lubricant films deposited out of a water carrier fluid can be much thinner while just as effective as the heavier oil films. In water-based emulsions, many constituents are immiscible in each other and the lubricant particles are generally negatively charged and held together in an electric field provided by carrier fluids and other additives. When these emulsions are applied to a die and the water carrier fluid evaporates, the resulting charged particles of lubricant orientate themselves on the die surface as a thin film to retain equilibrium between lubricant and additives. Solvent carriers, on the other hand, are non-polar and once evaporated there is nothing to neutralize or to balance the charges of the remaining lubricant particles; thus a thicker film results. In addition, solvent carriers are released slower than water and the material is often not completely removed from the die face before metal injection commences; this can lead to casting porosity and lubricant build up in the die cavity[2].

Die lubricant selection

The foremost requirement in selecting any die lubricant has to be the fast, easy ejection of a casting without health hazards. Once this requirement is satisfied, the quality of the die casting and acceptability of the lubricant under operating conditions can be taken into account, according to the considerations given in *Table 23.1*.

TABLE 23.1

Casting Quality	Production
Surface finish	Speed of casting
Ease of cavity fill	Ease of handling
Casting density	Maintenance
	Build-up on die face

No one lubricant meets all requirements for all die casters, and each lubricant has to be formulated for the type of metal being cast. An ideal lubricant at one temperature may cause problems at another, and with the limited capabilities of the lubricating ingredients available, it is unrealistic to expect an ideal universal lubricant.

The four main die casting alloys of zinc, aluminium, magnesium and brass are cast under different conditions of temperature and also display varying properties when they come into contact with the die lubricant and the die itself. At the two extremes of temperature, for zinc and copper alloys, the function of the lubricant in each case is different. In zinc die casting, the main concern is to lubricate the moving parts of the die and produce a surface finish suitable for any subsequent plating or painting treatments. Problems of casting release are not of great importance, since the shrinkage of the zinc alloy casting on to the cores is not appreciable due to the relatively low temperature. In die casting copper alloys, the high temperatures on the surface of dies and cores exert a significant effect. Lubrication of moving parts is essential and, with a high degree of casting shrinkage, lubricants must offer suitable properties to overcome these difficulties. Graphite suspensions in either solvent or water based lubricants are often employed. The soot deposited from an acetylene flame burning in air provides a thin adherent coating with excellent stability at high temperatures. It has been used successfully in copper alloy and ferrous die casting[3], as discussed on page 165.

Automatic die casting machines are enclosed by guarding, so water-based lubricants are recommended to avoid the fire risk of solvent-based fluids under those conditions. There is no doubt that modern water-based lubricants intelligently used have a parting ability equal to the solvent types in preventing intimate contact between die and metal. It is only in their ability to wet and spread spontaneously across the die and reduce friction on ejection that they are not equal. To overcome this it is necessary to select the most appropriate method of application. There are instances where a deeply recessed core may require the extra lubricating properties of solvent-based materials and in such cases they can be used selectively in conjunction with water-based compositions.

Having chosen a product for its correct application, easy ejection of a cast part is governed by the dilution rate; the amount of lubricant used when water-based materials are diluted 1:20 and sprayed for 5 seconds is the same as a dilution of 1:40 sprayed for 10 seconds. For a constant amount of applied lubricant, the ease of casting ejection would be increased with longer spraying time since more thorough coverage could be given. However, the disadvantage of slower production cycles must also be considered and an optimum concentration-spray time established.

Before changing to an alternative lubricant supply, both production rates and scrap levels should be monitored. A small percentage increase in scrap can often cost more than a year's supply of lubricant. If a new lubricant is being used it is necessary to experiment with the method of application to obtain the best results. Lubrication in pressure die casting is not just a function of the lubricant but also its method of application and the analysis in terms of efficiency, cost and performance must take both into account.

Application of die face lubricants

Three methods are employed to apply die face lubricants.

1. Hand spray guns.
2. Automatic fixed spray systems.
3. Automatic reciprocating spray systems.

A manually operated spray gun of the type illustrated in *Figure 23.2* is typical of the equipment which established confidence in the change from solvent to water-based die lubricants. The ratio of air to lubricant is normally controlled by valves attached to the gun. A carefully controlled spray pattern is important when applying water-based

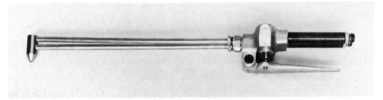

Figure 23.2 Hand spray gun which discharges lubricant and compressed air. (Courtesy Fenco – Aldridge (Barton) Ltd.)

lubricants. Unlike solvents which wet and spread over the die area dispersing the lubricant, once an emulsion contacts the hot die steel the water evaporates rapidly leaving a film of concentrated lubricant which is quite viscous and does not spread easily.

With the introduction of automatic die casting machines, it followed that automation of the lubricating cycle must be achieved. Several fixed nozzle spray systems were

Figure 23.3 Lubricant mist generator installed on a hot chamber die casting machine. (Courtesy Norgren Ltd., and EMB Co. Ltd.)

developed, some providing a mist spray suitable for small zinc die castings, others with high pressure nozzles providing more lubricant for aluminium and for large zinc dies. *Figure 23.3* illustrates a mist generator in which lubricant is atomized adjacent to the pressurized holding vessel and is dispersed either directly from open-ended tubing or through nozzles attached to the transfer tube.

For larger dies more lubricant needs to be applied in any one cycle than the mist systems can accommodate, and for cold chamber machines with dies running at high temperature the increased volume of lubricant must be applied under high pressures to penetrate any vapour barrier formed on the die face, due to the 'Leidenfrost' effect described on page 22. *Figure 23.4* illustrates a typical fixed automatic nozzle system where lubricant is fed under air pressure from the storage reservoir to the nozzle head. Compressed air applied to the nozzle at the appropriate time in the machine cycle both atomizes and directs the lubricant on to the die face. Nozzles are attached to the extremities of the die or machine plate by universal brackets and adjusted to suit a particular die lubrication requirement.

With larger and more complex dies, the use of a fixed spray system is limited, due to the interference of different spray patterns and the inability to apply sufficient amounts of lubricant into deep cavity recesses. In an alternative method several lubricating nozzles attached to a common manifold are reciprocated vertically between the dies as illustrated in *Figure 23.5*. In this way the nozzles can be adjusted to avoid spray interference and, being in close proximity to the die area, can be directed to disperse lubricant into recesses not accessible from the extremities of the die.

Several types of extractor arms that reach into the die area at each cycle to retrieve the casting are employed and die spraying nozzles are often incorporated as an extension

Figure 23.4 Fixed nozzle system for die spraying. (Courtesy Acheson Industries (Europe) Ltd.)

Figure 23.5 Vertical die spray reciprocating system mounted on a cold chamber die casting machine. (Courtesy Acheson Industries (Europe) Ltd.)

to the arm. Whilst the arm is retracting with a casting firmly gripped, a spraying sequence for the next cycle is completed. This can lead to increased speed of casting, since the reciprocating movement for extraction and lubrication has only to be carried out once. With all types of spray lubricating equipment, air can be used to clean the die surfaces before lubrication and often a combination of solvent based and water based lubricants can be applied together to selected die areas. The provision of adequate safety interlocks to prevent damage to spray equipment and die as the machine closes the two die halves is, of course, essential.

Electrostatic equipment is a novelty technique for applying lubricants but it is unlikely to become popular in die casting for several reasons. The thin lubricant film required in die casting is not sufficient to insulate electrically the steel surface, and excessive lubricant build-up can occur. Water-based lubricants are conductive and their electrostatic application requires additional expensive equipment. On the hotter die areas a steam layer will repel the spray and charged lubricant droplets are deposited on the colder die areas, where they are not needed.

With increased application of improved water-based lubricants, many companies are employing centralized mixing, storage and distribution. This has become possible because these types of lubricant are stable; there are no fire hazards and often a fixed dilution rate will be adequate for a wide range of castings. Automatic systems are available which can be used either to prepare bulk supplies or meter small proportioned amounts at each machine. Agitation using mechanical paddles or methods employing recirculation piping, where the return flow is connected to the bottom of the tank, are suitable. Some manufacturers advise against agitation of those lubricants which are susceptible to scum formation. Dispensing lubricant from a central storage system can be achieved either by a closed loop air pressure system or, more usually, by positive displacement where gear type pumps are used. Centrifugal pumps are not recommended for water-based lubricants since their shearing action can cause instability and breakdown of the emulsion.

Plunger lubricants

The injection plunger and cylinder on a hot chamber machine is immersed in a low melting point alloy and, because pressures and plunger speeds are low compared to those of cold chamber operation, lubrication in these areas is not required. With cold chamber machines using horizontal or vertical injection systems, requiring pressures up to 30 000 lb/in^2 (200N/mm^2), close dimensional plunger tolerances and involving severe frictional problems, efficient lubrication is necessary if galling and wear are to be minimized. Frictional forces become extremely high in the final stage of plunger travel and the resulting wear of plunger and sleeve is related to the combined effect of heat and pressure generated during the shot. Plunger lubricants are used to reduce the frictional and shear forces encountered at high temperatures and are usually based on greases and heavy oils incorporating graphite.

The oils and greases themselves probably do not contribute to the performance, but act simply as carriers for the additives and pigments. This phenomenon appears to be confirmed by recent work in Japan where in conjunction with the pore-free die casting process, discussed on pages 235–239, dry films of boundary lubricant have been used successfully, although an overriding problem was the difficulty in applying dry film lubrication. One relatively new solution to the problem has been the application of water-based plunger lubricants; with correct application, these provide superior service. Reduced casting staining, porosity and smoke emission are reported but since water evaporates faster than oil, a film of relatively viscous lubricant material is deposited and may not be carried along the sleeve. Correct application is, therefore, essential to deposit the lubricant as far as possible down the injection sleeve to the area where the greatest friction and wear occur. A paper by A.W. Cooper[4] reported on the build up of frictional forces during plunger travel and compared the performace, properties and methods of application of various plunger lubricants.

Plunger lubricants are applied either manually, by brushing on to the plunger and sleeve, or dispensed automatically from a pump initiated by an impulse from the die casting machine as the plunger is in the retracted position. *Figure 23.6* shows a pump where lubricant is dispensed directly onto the plunger or injected into retaining grooves either within the sleeve or to a separate split collar attached to the plunger sleeve.

Figure 23.6 Electrically operated plunger lubrication pump. (Courtesy of Kluber Lubrication GmbH)

Die casting machine lubrication

Trouble-free operation of all moving parts of a machine must be maintained. Lubricants for this purpose may be of either oil or grease formulation, incorporating extreme additives giving thermal stability and shear resistance. Moving parts such as tie bar bearings, knuckle joints in the locking systems and bearing plates on which the platens move, operate under very high pressures and loads, thus regular lubrication is required. Various types of equipment are available for dispensing small metered amounts of suitable lubricant to these machine parts. Lubricant in a central reservoir can be pumped at predetermined intervals into a line system having various connections to the machine's bearing surfaces. Alternatively, each line terminates with an individual pump unit but in either case a continuous economic and controlled injection of lubricant is the essential feature of the system.

Safety and stability of lubricants

With the advent of water-based die face lubricants fire hazards and pollution have been reduced. After the water has evaporated, the residual lubricant normally has a high flash point and is deposited as a thin film which precludes any formation of combustible vapours. Much effort is being directed to the efficient collection and removal of lubrication fume and this is made easier when water-based lubricants are used. Toxic and skin irritant constituents must be avoided; current legislations are in force to control the use of such materials, these being amended as more experience is accumulated. Disposal of lubricating products from washing of storage tanks, etc. is also controlled. Local Authorities must be consulted to ensure that discharges comply with standard recommendations. Stability requirements of a lubricating product normally refer to a particular lubricant's property before its application. These may include storage stability, hard water compatibility and corrosion and bacteria.

Storage stability

Lubricant emulsions can be broken down to yield their constituent fluids by freezing, heating, pressure and agitation. It is important, therefore, that stability is maintained whilst the lubricant is in store so that consistent properties are achieved during use. Any pigments present should also be stable to assist subsequent blending with solvents.

Hard water compatability

Compatability of a lubricant with hard water can avoid the formation of any scum precipitates which could otherwise cause blockage of spray equipment.

Corrosion and bacteria

Lubricants should be formulated to prevent corrosion of storage drums, spraying equipment, dies and die cast parts. In addition to use of corrosion inhibitors, chemicals to provide resistance to bacterial degradation are often specified. Simple tests may be conducted to ensure that these properties are adequate for the lubricant's intended use. Build up of a die face lubricant can occur on a die, resulting in unacceptable casting quality. This is caused by the accumulation of die lubricant ingredients which turn into resinous solid matter — waxes are particularly noted for this tendency, even though they have good casting release properties. Care in maintaining die temperatures and in the selection of a lubricant can help to reduce this problem, although use of caustic cleaners and mechanical polishing may eventually be required.

Future developments

Many advances have been made from the beginnings of lubrication technology but some aspects are not yet fully understood. Tribology centres have been set up recently to study in more depth the complex sciences which govern lubricant formulation and application. In Britain the Industrial Unit of Tribology at Leeds University has made valuable contributions in many industries but so far their interest in pressure die casting has been limited. They publish an interesting journal 'Industrial Lubrication and Tribology', at present appearing six times each year.

As discussed on page 220 polymerization of silicones leads to hard deposits on a die surface, but silicone fluids are now being developed based on the use of cyclic silicone compounds which, because each molecule chain has no beginning or end, is unable to polymerize.

Natural waxes, often used in the early days of die casting, provided effective release properties but, due to their tendency to decompose, were detrimental to the appearance of the casting. Development of synthetic waxes which exhibit lubricating and separating properties similar to those of the natural waxes, but which are chemically more stable, are being used to an increasing extent. During the past 20 years lubricants have evolved from about 10% oil in solvent to water-based lubricants containing less than 0.2% solids. The future trend will continue in the direction of new lubricants which can be expected to be as cost-effective as the ones now available while providing better casting release, faster production rates and higher quality castings.

Lubricant manufacturers have always been aware that the pressure die caster's dream is to eliminate consumable die lubricants and incorporate a permanent release coating on the die surface. The production of porous surfaces on a die face to accept a permanently bonded solid lubricant has been investigated. Mixtures of graphite and molybdenum disulphide have been bonded to dies by the use of organic and inorganic bonding agents but at present the self-lubricating surfaces which have been produced are not sufficiently reliable to be used without supplementary fluid lubrication. Some tentative steps have been taken in plastic injection moulding but even there the use of permanent coatings has had only limited success and the disadvantages of having to clean and recoat after a period of time outweighs any advantages gained. As far as the Authors are aware the production of a self-lubricating ejector pin has still not been achieved satisfactorily but the prospect offers so many advantages that the problem is bound to be solved sooner or later. Should the use of permanent coatings be achieved, process variables with hand or mechanized application of consumable die face and die movement lubricant compositions will be eliminated.

References

1.   SCHMIDT, A.B.
      Mechanism of die lubrication in aluminium die casting. *Die Casting Engineer,* **24,** (3), 14 (May/June 1980) (Also printed in *Foundry Trade Journal, Die Casting Technology Supplement,* **150,** (3209), 622–632 (March 26th, 1981)

2.   VAN WINKE, E.F.
      Die Casting release and lubrication technology for water and solvent-carried compound systems. *Society of Die Casting Engineers* Congress Paper No. 1302 (1966)

3.   MAIER, R.D.; WALLACE, J.F.
      Die Casting of copper alloys. *Transactions of the American Foundryman's Society,* **81,** 194 (1973)

4.   COOPER, A.W.
      Current Trends in plunger lubrication. *Society of Die Casting Engineers* Congress Paper No. G-T77-046 (1977)

Chapter 24

New developments in die casting

Computer-aided die design

The impressive results of thin-wall die casting techniques described on page 103 have been reinforced by the 'new technology' which is now available to help Managements to design dies scientifically and rapidly[1]. A comparatively inexpensive computer provides a data base which instructs what parameters should be selected to ensure the best possible conditions for production of a given die casting. The first system for the small desk top micro computers was an ILZRO–Battelle–Zinc Institute (USA) program produced in 1979 for the Tandy TRS80. In 1980, programs were brought out by Mazak Limited for zinc alloy die casting companies and by BNF Metals Technology Centre for gating and thermal die design of zinc, aluminium, magnesium and brass die castings. The systems differ in scope and type of computer but have similar purposes.

The MAZAK and BNF gating programs require information to be stored on the injection performance of the machine. Much of this information came from a study of machine injection characteristics referred to on page 104 in connection with zinc thin-wall die casting. At first, the information was expressed in the form of graphs but the development became more acceptable to die casting managements when the information was programmed on computers, the Tandy TRS80 for the MAZAK system and the Commodore Pet for the BNF Gateway system.

Each computer program can be tailor made for the individual die casting company. Pressure and metal flow rates are stored as equations; the designer decides which machine is to be used and feeds to the computer relevant details including the weight of the casting. The computer calculates the range of combinations of gate area, metal velocity and cavity fill time and the information is displayed on the screen. Once the gate area is selected the computer continues, calculating the acceptable range of nozzle or plunger tip sizes. The whole operation is conducted as a series of logical question-answer stages, leading to the selection of the cross section areas of each part of the runner system and the configuration of runner and gate to give smooth metal flow.

The Gateway system provides data on a large number of die casting machines; by late 1980 information has been programmed on about 20 machines, and the list is being enlarged as further trials are performed. The system has been extended to cover costing and weight calculations. A thermal program[2] is available which calculates the heat input to the die by the die casting alloy and estimates what cooling capacity is required and optimum size and position of cooling channels.

The benefits of the new technology include improved productivity, reduced casting weight and reduced shot weight, leading to reduced melting cost. The casting parameters and gating design can be determined logically and rapidly, but the new technology has brought other benefits. A company experienced in the use of the new systems can ascertain whether a projected new component can be produced effectively on existing machines. Before quoting for a new proposition the computer has checked that the casting will be successfully produced and if it 'advises' to the contrary the management can either investigate whether the conditions offered by a more powerful machine will make the casting successfully, or they do not quote for the proposition. The element of guesswork based on experience has been removed; die casting companies, on receipt of an inquiry, can either report that the component is not suitable for their plant, or, having confirmed that it will be suitable, have benefited by the status gained in this way.

Vacuum die casting

The pressure exerted on the injected metal is several hundred times greater than that of the air in the cavity, so any air entrapped in the die will be compressed into microscopic bubbles. During a study of the process in 1961, F.C. Bennett[3] calculated the changes in air volume in a die cavity with different metal injection pressures assuming that none of the cavity air was vented. Pressures typical of those used in hot chamber die casting showed that the volume of air would be reduced to about 0.75% of the cavity or casting volume, while at the higher pressures used in cold chamber production it would be reduced to about 0.45%. In addition, the cold chamber machine injects extra air from the space over the metal charge in the sleeve and this causes a further increase of the air volume. Since these calculations were in general agreement with practical experience it was questioned whether normal die venting techniques were fully effective or whether lubricating vapours evolved during metal injection maintained the die cavity full of gas. Bennett's figures, however, represent the possible gain to be achieved if a die cavity is evacuated before metal injection.

Attempts to develop methods of vacuum die casting had been directed towards the casting of the low melting point alloys in hot chamber machines as early as 1915, when C.M. Grey[4] patented an air injection machine with some similarity to a contemporary hot chamber vacuum die casting unit. In 1941 Brunner and Trebes[5] patented a machine which featured a mould evacuated and sealed by means of external flanges at the parting plane. During the 1960s interest in the possibility of vacuum die casting was world wide. A series of articles by H.K. Walker was published in the British journal *Metal Industry* in 1962 and summarized in an information bulletin[6] by the Society of Die Casting Engineers in 1963. This gave a remarkably comprehensive survey of the various methods which were being used, and the problems which were arising from the use of those methods.

Later developments led to the increased use of vacuum die casting with major advances in the field of cold chamber aluminium production. As the entrapped air is compressed in conventional die casting, a back pressure is generated within the die cavity which increases rapidly as the venting areas are sealed by metal. Metal flow is subsequently retarded towards the end of the die filling and parts of the casting formed by the metal first injected may begin to solidify before the final intensification pressure is effective. Where isolated bosses or other mass concentrations are present in the casting they will not benefit from the full effect of pressure intensification so porosity may be present in these areas. An important advantage of vacuum die casting is that by reducing back pressure, a faster rate of metal flow allows the intensification pressure to be applied before local solidification begins. Although a high degree of solidity can be achieved in 'non-vacuum' die casting by good gating and venting techniques, the higher die filling rates achieved with vacuum makes possible the production of large casting areas with thin sections. Although this gives a wider scope to the product designer, skill and experience in the field of die design is demanded even more for vacuum than for conventional die casting.

Although it had been obvious for many years that vacuum die casting would provide a method of obtaining pressure die castings free from microporosity, the efforts to perfect a system that would be compact, effective and economic have been extremely difficult and did not come to fruition until the 1970s.

During the 1960s, vacuum die casting developments included the introduction of systems which employed evacuable hoods where either the entire small hot chamber machines were enclosed in the hood or, more often, the die area of the machine was enclosed. These techniques such as that of the 'Nelmor' system required a vacuum capacity many times greater than the hood itself for satisfactory pressure reduction to be achieved in the shortest possible time. Large surge tanks were used which were repeatedly evacuated by a vacuum pump. One attraction of the hood system was that existing dies did not require modification, although the greatest limitation was on the space available for core withdrawal mechanisms and other projections from the die. Closer fitting hoods around the die were offered with gasketed panels which could be used for any projecting die parts. Although in principle the idea of enclosing the whole die with a hood may have appeared logical, the apparatus was cumbersome, the amount of air to be removed at each shot was very considerable and this process has been practically discontinued.

Before a vacuum can be achieved in cold chamber machines it is necessary to shut off the pouring hole in the injection sleeve after metal ladling. The simplest method is to advance the plunger beyond the pouring hole and dwell at this position until the required level of cavity evacuation is achieved. Various pouring hole covering devices are also available, some of which incorporate separate vacuum lines.

Several 'hoodless' systems have been developed which evacuate the die after closure and maintain the vacuum until the die cavity is filled with metal. One method of evacuating air from the die cavity is by way of the ejectors, the ejector box acting as an intermediate vacuum chamber which relies upon the movement of the ejector plate to cut off the metal flow along the vacuum channels. Another method is to evacuate air at a point along the die edge; a separate scavenging vacuum positioned around the cavity form helps to prevent inward leakage of air reaching the cavity.

Another method used for vacuum and metal cut off at the end of each cycle is the DCRF chill plug for massive venting. Vacuum channels are cut in the face of the die as required and extended to the die edge on which is mounted two water cooled blocks with abutting faces. A broad, shallow ridged or corrugated air passage formed between the blocks gives a 'washboard' effect to encourage rapid chilling and stop the flow of metal before it reaches the top of the blocks from where the vacuum line is taken. Although the

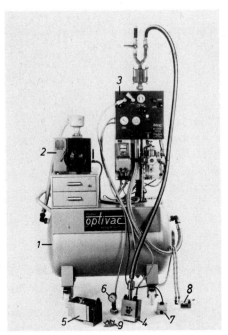

Figure 24.1 Complete vacuum installation. 1 = vacuum tank; 2 = vacuum pump; 3 = automatic 'Mecanovac' control unit; 4 = superstop 'Mecano' (evacuation valve); 5 = oscillostop 'Mecano' with integrated shockabsorber and compensator; 6 = vacuum control indicator; 7 = vacuum signal lamp; 8 = vacustart valve (pneumatic); and 9 = 'Mecanovac' symbol. (Courtesy Fondarex, F. Holder & Cie)

device has been used with vacuum assistance, reduced casting porosity has been reported when the blocks are just open vented to atmosphere. This effect has been frequently observed on dies where extra venting is provided when designing for vacuum.

The Hodler system supplied by the Swiss firm Fondarex, F. Hodler & Cie was initially developed and patented during the early 1950s and has been continually improved over the years until at present it is widely accepted in many parts of the world, about 450 systems having been installed to the present time. *Figure 24.1* shows the 'Optivac' (optimum vacuum) system with the recently modified 'Mecanovac' Control Unit[7]. Immediately after the die casting machine's injection plunger has moved past the shot sleeve pouring hole, the evacuation valve is opened following an impulse from the 'Vacustart' valve; the die cavity is evacuated through a section channel cut in the die and connected to the casting at the point where the die is last filled with metal. One distinguishing feature of the 'Mecanovac' is the manner in which the impulse is given for closing the evacuation valve assembly attached to the die. It is not dependent upon the time or movement of any of the electrical, hydraulic or mechanical operations of the die casting machine, but solely upon the hydraulic force of the liquid metal stream[8]. When the die is completely filled the evacuation valve closes under normal operating conditions in 1/800 second.

Improvements in this system include the mechanical means by which the evacuation valve closes upon arrival of the stream of molten metal and offers improved reliability over the auxiliary power systems previously used. Increased evacuation capacity of the Mecanovac valves enables the production area of the vacuum runners in the evacuation valve to be reduced by more than 40% and the mounting and removal on the die is simplified. The Hodler system has the advantage that it is both developed and operated by a reputable die casting company. The major applications of this system have been for the production of high quality die cast components that require pressure tightness, high mechanical strength or which are to be heat treated. Heavy and complex castings such as transmission housings and engine blocks are being produced with the aid of Hodler vacuum technique. Although this method can be applied to a wide range of casting sizes, it appears that at present die castings weighing 0.5 kg upwards are suitable. The application of vacuum permits lower pressures and therefore smaller machines can be used.

Like most technical developments, vacuum die casting is not a 'cure all'. A die casting which is producing 20% scrap might indeed be improved by the application of a vacuum system, but first the casting and die design and the arrangement of the gating system should be investigated. Vacuum die casting is not intended to cure substandard production but to improve the quality and solidity of die castings that would normally be commercially acceptable but which, for reasons of severe stressing, heat treatment or exceptional requirements, need to be of an even higher quality.

Pore-free die casting

Throughout the history of pressure die casting it has been acknowledged that the rapid injection of metal under pressure into an air filled die cavity causes a certain amount of porosity. Venting to allow air to escape, follow-up increased pressure immediately after the metal was injected and the recent developments in the scientific gating of dies all helped to limit the amount of porosity but it had to be admitted that most pressure die castings were not suitable for heat-treatment, since the furnace temperature causes even the smallest areas of microporosity to swell and cause blistering. Highly stressed components such as automobile wheels are, indeed, pressure die cast successfully, but it is necessary to monitor their production by X-ray examination to ensure that any porosity is situated where it will not weaken the wheel in critical positions. In another field, pressure die cast lead alloys can only be used in limited areas in battery manufacture because even a minute patch of microporosity would introduce the hazard of electrolyte leaking. Consequently, automated gravity die casting is often preferred for battery grids and terminals, because that process is more likely to produce porosity-free castings.

In the early 1970s technicians in Japan, the USA and the USSR were responsible for developing a new approach to the production of sound die castings. Since oxygen is the agent which reduced the porosity, the process was at first called 'oxygen purging' but, perhaps because the word 'purging' is emotive, the designation 'pore-free die casting' has now become more widely used. The original concept of pore-free die casting came from a Canadian research on the composition of gases trapped in the pores of zinc pressure die castings[9]. The gases consisted mainly of nitrogen from the air and hydrogen from the die lubricant hydrocarbons; however no oxygen was found, although air contains about one fifth oxygen. From a further study of these results it was concluded that the oxygen could only have disappeared by reacting with the zinc to form oxide which then dispersed in the metal. From this it was reasoned that if the die cavity was purged with pure oxygen before metal injection, all other gases would be driven out and the oxygen would react with the metal, forming oxide, and leading to a porosity-free casting. Additional researches were performed to study the thermodynamics of the reaction and to confirm that no explosion danger existed. In fact the consumption of oxygen in the process causes a rapid pressure drop and a partial vacuum condition, which improves the flow of metal into deep cavities. This work resulted in the issue of US patent 3 382 910 to the International Lead Zinc Research Organization.

While the American research was developing, similar work was being done in Russia and Japan, the latter under the direction of Nippon Light Metal Company Ltd. Later ILZRO and Nippon concluded a cross-licensing agreement exchange of information and promotion of pore-free die casting technology[10]. At present there are 11 commercial licensees in Japan and one in the USA. The major expansion of the pore-free process has been in Japan, in particular for those sections of industry supplying aluminium pressure die castings for automobiles and motor cycles. In the USA, the ILZRO developments have been principally in the field of zinc and lead alloy die casting.

In 1970, Fuso Light Alloys Ltd., now the principal producer of pore-free die castings in Japan, launched joint studies of the process with Nippon Light Metal Co. Ltd. In 1972 they initiated volume production of heat-treated motor cycle wheel hubs, and by 1980 their output amounted to 1500 tonnes. The principal uses of pore-free die castings include parts that require high strength, such as connecting rods and wheel hubs. Die castings that must resist heat, such as hot plates for home cooking, are also made by the pore-free process.

An early application related to automobile air conditioners. These were made previously by conventional pressure die casting but, even after impregnation, the cylinders suffered a high rate of rejection. Leakage occurred from drilled holes and around cast-in inserts. When pore-free die castings were used, rejection rates dropped substantially and the castings did not require impregnation. *Figure 24.2* shows two car cooler cylinders and a motor cycle hub.

When molten metal is injected into the oxygen-filled die cavity, the oxygen reacts with the aluminium, forming very finely dispersed aluminium oxide, thus eliminating gas bubbles in the metal. The oxide particles are less than $1\,\mu m$ in size and no adverse effects are noticed. The process is suitable for hot and cold chamber production, and for die cast

Figure 24.2 Car cooler cylinders and a motor cycle hub produced by the pore-free process. (Courtesy Fuso Light Alloys Ltd.)

alloys of aluminium, zinc and lead. So far no significant developments have been shown for magnesium alloys, but research in that field is continuing.

Slight modifications to the die casting machine are necessary, involving halting the plunger at the end of the slow approach. A valve is then opened, allowing oxygen to be introduced into the die. The oxygen flows for a preset time and then the fast shot is energized. It is necessary for safety reasons to use only water-based die lubrication. In the cold chamber process, the sleeve and plunger must be lubricated by means of a proprietary system which is not necessary for hot chamber machines. The volume of the cavities that must be flushed by oxygen in the two processes is different. With cold chamber equipment both the die cavity and the sleeve must be flushed and the sleeve pour hole must be sealed for effective flushing of the die cavity. This operation also involves proprietary hardware. In the hot chamber process a two phase shot (a slow and fast metal displacement rate by the immersed metal piston pump) is used, and the injection system is modified so that the plunger stops at the end of the slow shot. Plunger movement during the slow shot seals off the pump inlet holes and displaces the metal to the end of the gooseneck nozzle (the end of the hot metal passages between the pump and the die). The only area that then needs to be flushed is the die cavity itself. Oxygen is introduced into the cavity through a channel thin enough to freeze off and prevent the metal from flowing into it during injection. Stopping the plunger automatically starts the oxygen flow into the die cavity (for up to several seconds), flushing the cavity gases out through the cavity vents. The actual amount of oxygen used depends on the size of the cavity. The oxygen flow is controlled by solenoid valves; the timer signals to stop the oxygen flow, activates the fast shot and the rest of the cycle then proceeds in a normal fashion.

In the development of the process for the hot chamber die casting of zinc, a series of trials were carried out. Improvements with pore-free zinc die casting were shown over a range of casting conditions involving variations in gating, venting and metal injection conditions. These are important in maximizing cavity purging and consumption of the oxygen. The presence of oxides in the castings has not proved to be detrimental, since they are very small and finely dispersed in the cast structure. A common method of determining the amount of residual gas porosity is to subject the zinc castings to a blister test of 2 hours at $350°C$ and then examine them visually. *Table 24.1* shows the density of aluminium alloy die castings, before and after heat treatment. The first figures are for castings made without oxygen purging. During heat treatment the microporosity that was present became enlarged so that the density of the casting was reduced. The next three lines show the effect of increasing times of purging. With 4 seconds of oxygen treatment the castings were so solid that after heat treatment they were little changed in density. The amount of aluminium oxide was measured in each casting.

TABLE 24.1

| Oxygen treatment (s) | Density | | Al_2O_3 content (%) |
	before heat-treatment	after heat-treatment	
Without oxygen	2.621	2.371	0.080
2	2.660	2.582	0.195
3	2.656	2.584	0.200
4	2.658	2.634	0.240

In addition to aluminium and zinc there are great possibilities for die casting lead alloy battery components which must be pore-free, such as grids, terminals and intercell connectors. *Figure 24.3* shows sections of a top terminal bushing for a lead-acid battery produced with and without oxygen.

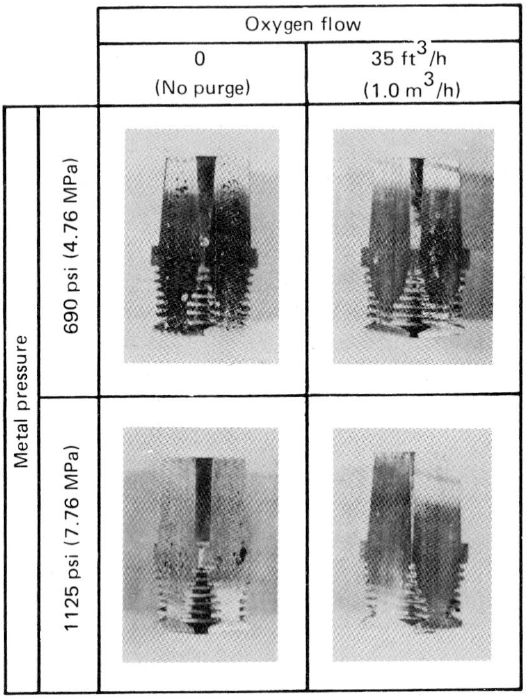

Effects of oxygen flow and metal
pressure on internal soundness

Figure 24.3 Top terminal bushings for a lead-acid battery produced with and without oxygen. (Courtesy ILZRO)

In conventional die casting high injection forces are necessary, partly to compress gas pores, but, when the porosity is eliminated, the casting pressure can be reduced and thus a machine of lower tonnage can produce equally good results. In a paper on pore-free die casting by Usao Miki and Takeshi Kido[11], an example was quoted of an aluminium alloy cooking pot with a casting area of about $600 \, cm^2$ which had previously been die cast on a 500 tonne machine. After the die had been adapted for pore-free production it became possible to cast the pot on a 250 tonne machine.

One of the principal outlets for pore-free die castings is the automobile wheel. The severe conditions of service of such wheels demands freedom from porosity, high strength and good ductility under impact conditions. Although the pore-free process permits the heat treatment of castings produced by it, the normally available aluminium alloys are still deficient with regard to ductility. On this account Nippon Light Metal Company Ltd. embarked on a program to develop an alloy that would provide great ductility. The research resulted in a new aluminium alloy designated as DX-30 containing 8.0-10.0% silicon, 0.30–0.35% magnesium, 0.6–1.0% iron and 0.2–0.5% manganese. The manganese addition has the effect of spherodizing the Al–Fe–Si ternary compound, leading to an improvement in ductility after solution heat treatment.

The major production of pore–free die castings has been concentrated mainly on aluminium, in view of its developing use in automotive fields and the definite interest in obtaining an aluminium alloy die casting that can be heat treated. However, great interest has been shown in the possibility of making zinc alloy die castings free from subsurface porosity, to reduce the cost and improve the quality of plated die castings. Reports on the application of the pore-free process to zinc die casting have been given to the Society of Die Casting Engineers in 1977 and in 1979 by Louis Battiston[12,13].

Zinc alloy die cast plates 15 cm x 7cm were tested; the thickness could be adjusted and the gating modified. A blister test was used to evaluate the solidity of castings, with varying gate configurations produced with and without oxygen. The solidity of fan-gated castings was only slightly improved but, with single or triple slit gates, designed for pore-free production, a significant reduction in porosity was achieved.

The casting cycle rate is diminished by about 10% due to the time required to flush the die and the flow time is dependent on the shot volume. Oxygen has to be provided costing a few percent of the raw material metal cost. In addition the royalty cost of operating the patented process has to be added. According to a report by S.F. Radtke[14], the pore-free method will result in a cost increase of 10–15%. The savings to be expected include reduction in amount of scrap and reduced expenditure on quality control. The main advantage, however, is that castings that previously were outside the scope of pressure die casting can now be produced successfully.

Die casting semisolid metals

Within the last 10 years a new family of processes have been developed, emanating from the understanding of the behaviour of solidifying alloys. In combination with existing technology, the new processes have given birth to a more efficient use of energy, materials and machinery.

Alloys for conventional die casting are melted to temperatures above their liquidus, then measured quantities are injected into the die cavity. Superheat is needed so that the bulk of solidification occurs only when the cavity is full, since a volume fraction of solid content in excess of 0.2 (that is when it is more than 20% solid) causes the alloy to stiffen and resist flow. This occurs since conventionally solidified alloys have branched dendrites, which develop interconnected networks as the temperature is reduced and the volume fraction of solid increases. In contrast, the new processes involve the casting of specially prepared alloys at lower temperatures, intermediate between liquidus and solidus, so that their superheat and about half their heat of fusion have already been extracted before casting. In the past it has been considered that the greatest advantages are to be obtained in the production of die castings from relatively high melting point alloys, including copper-base and ferrous metals. However, more recently research efforts have been concentrated on aluminium alloy die casting, where it has become apparent that the process will have great potential and, during the next few years, more commercial production by the semisolid processing techniques can be expected.

Advantages of the semi solid processing techniques are listed below.

1. Short solidification time — no superheat and reduced latent heat, hence shorter cycle times and lower energy costs.
2. Lower heat input to dies — less thermal shock, hence longer die and shot sleeve life.
3. Viscous fluid state gives non-turbulent die filling with no flash or air entrapment — thick section components can be die cast in heat treatable alloys.
4. Reduced solidification shrinkage — improved casting tolerances and surface definition. Lower machine locking forces — final pressure intensification is not required.
5. Accurately metered shot weights — eliminating the need for conventional holding furnaces, thus saving metal and energy.
6. Alloys readily flow into thin sections.
7. Improved safety and working environment.

The possibilities of die casting with slurries of semisolid metal was an unexpected outcome of discoveries made whilst studying the flow properties of partially solidified alloys at the Massachutsetts Institute of Technology in 1970. D.B. Spencer, a former doctoral student working with Professor M.C. Flemings and Professor R. Mehrabian discovered that when tin-lead alloys are vigorously agitated during solidification a very different structure is observed[15] from the usual interconnected dendritic structure. The primary solid particles comprise degenerate dendrites or nodules which are spheroidal in shape, separated from each other and suspended in the liquid matrix at solid levels up to 80%. Agitation prevents the particles from coming into contact for any length of time, resulting in a semisolid slurry having thixotropic properties (their viscosity decreasing with increasing rate of shear, is time dependent and reversible). The primary solid particles are made up of one or more phases having an average composition different from that of the surrounding matrix. Work has been extended to slurries of aluminium and zinc alloys, bronzes and more recently stainless steels and a cobalt-based alloy.

In the UK, the Fulmer Research Institute at Stoke Poges have been applying semi-solid processing techniques to aluminium alloys using cold chamber die casting machines up to 500 tonne locking force. *Figure 24.4* shows the microstructure of an LM21 alloy prior to die casting having a 60% volume fraction of solid. The three important process variables affecting the structure of a slurry are average shear and cooling rates and volume fraction solid. Viscosity depends on how much solid is present and, with reduced agitation, the mixture thickens considerably as the particles are allowed to contact and join together to stiffen the structure. If agitation is stopped when the alloy is about 50% solid, its apparent viscosity rises rapidly even though the solid content remains constant and the semisolid slurry can be handled in a pair of tongs as a soft solid. From this stage the slurry can be reconverted to its liquid form by applying forces to shear all the bonds, viscosity reducing to such an extent as the shear rate increases that the alloy flows again.

Extending this principle to pressure die casting, the slurry charge prior to injection behaves as a solid but shearing, which occurs at the gate entry during injection, reduces the viscosity of the slurry to a level at which it flows smoothly. Die filling can proceed with lamellar flow at relatively high gate velocities without the excessive turbulence and splashing that results when low viscosity metal is 'sprayed' into a die in conventional die casting.

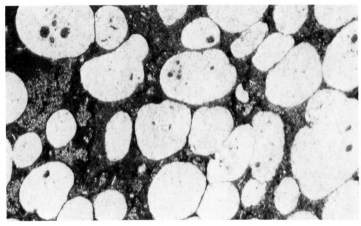

Figure 24.4 Microstructure of semisolid LM21 (319.1) alloy showing white spheroidal particles of aluminium rich phase in the eutectic matrix (100X). (Courtesy Fulmer Research Institute)

Castings produced from metal slurries are variously known as rheocastings, thixocastings or compocastings according to the particular processing technique or material composition employed. *Figure 24.5* illustrates the principles of the semisolid casting processes. The term rheocasting describes the most direct sequence for the production of die castings, and involves transfer of the rheocast slurry into the injection sleeve of a cold chamber pressure die casting machine and injection into the die cavity. Thixocasting refers to the reheating of solidified rheocast slugs into the liquid-solid range, subsequent to charging into the injection sleeve. Compocasting involves the introduction and retention of fibrous or particulate non-metallic filler in the slurry.

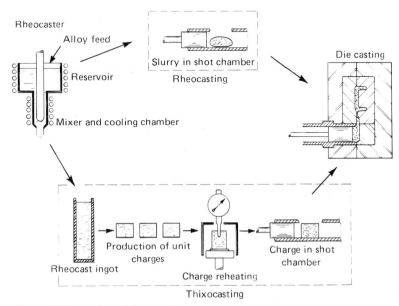

Figure 24.5 Principles of slurry casting process

Rheocasting

The basic equipment required for each casting route consists of a 'rheocaster' which originally comprised a simple paddle stirring system. The need for more controlled casting conditions led to its replacement by the rheocaster, shown in *Figure 24.6,* having two vertically connected concentric cylinders. An upper reservoir chamber holds the molten alloy, and below this, a smaller diameter chamber acts as a heat exchanger and agitation zone. Shearing is provided by a rotating graphite or ceramic rotor which passes through the reservoir and mixing chamber forming the exit nozzle for the slurry.

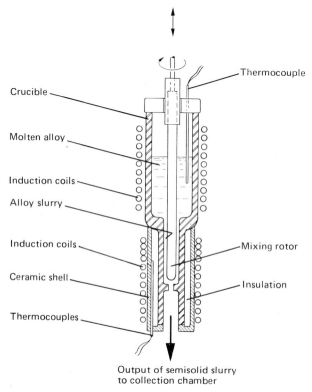

Figure 24.6 Schematic diagram of high temperature rheocaster

Graphite has been used successfully for the rotor and chamber lining for producing yellow metal slurries but, for ferrous alloys, recrystallized alumina has proved to be a satisfactory material giving long service life. Rheocast slurry is produced continuously allowing alloy to flow through the mixing chamber, where it is simultaneously sheared by the action of the rotor in the small annular space, and cooled through the mixing tube walls by the surrounding water cooled copper coils. For a given rate of heat extraction through the crucible walls (a function of crucible conductivity and cooling coil/crucible separation), increasing the flow rate reduces the time the alloy is in the mixing chamber and decreases the volume fraction solid of the slurry. Lowering the flow rate has the opposite effect. Design of rheocasting equipment must ensure that cooling rate is slow enough to permit formation of a spheroidal non-dendritic structure. Provided the shear rate is sufficient to obtain this structure, increased cooling rates will reduce the particle size.

Mechanical rheocasting methods suffer from several inherent problems. In addition to erosion of the rotor, the small annulus formed between rotor and chamber wall gives low volumetric flow rates and for commercial application it becomes difficult to couple mechanical agitation to a continuous casting system. Flow lines are likely to be present in the rheocast alloy due to interrupted flow and/or discontinuous solidification, and slurries with more than 60% solids are difficult to achieve since the rotor is immersed in the slurry. To reduce these problems, ITT Industries in the USA has recently proposed electromagnetic stirring described in several patents. Whilst this technique is an improvement over mechanical methods, there are some limitations imposed by the nature of the stirring. It was found that a stirring motion associated with a rotating magnetic field generated by a conventional two pole three phase inductor motor stator achieved the required shear rates for producing thixotropic semisolid slurries with uniform structure. The loss of magnetic field strength is small due to the low frequency that is used and there is very little resistance heating of the melt being stirred.

In operation, molten metal is poured into the top of the mould while the motor stator, arranged circumferentially about the mould, causes the molten metal and slurry to rotate. Slurry is withdrawn from the bottom of the mould in a continuous or semi-continuous manner, as in the mechanical rotor system. The mould is formed from any non-magnetic material such as stainless steel, copper or copper alloy. An insulating ceramic liner is placed inside the upper region of the mould to postpone any normal dendritic solidification before the molten metal is in the region of the magnetic stirring force. It is reported[16] that, using this principle, molten aluminium alloy cooled at an average rate of 50°C per minute employing a rotating magnetic field generated with a current of 15 amp produced rheocast material over 6 cm diameter at speeds up to 36 cm/minute.

Thixocasting

The alternative casting method, known as 'thixocasting' from the thixotropic properties of the slurry, allows the output of the rheocaster to solidify in a suitably shaped mould, after which it can be reheated into the semisolid temperature range before introduction to the shot sleeve of the cold chamber machine. Rheocast ingots cut into the appropriate sizes for each casting shot are reheated without agitation in an induction furnace. The softness of the charge is continuously monitored using a weighted silica or alumina probe indicator which rests on the charge and penetrates the slurry progressively. The desired volume fraction of solid is indicated by mean velocity and displacement. The cylindrical semisolid charge is transferred to the shot sleeve; the shearing which occurs at the gate entry during injection reduces the viscosity of the slurry to a level at which it flows smoothly into the die. The die caster has the advantage in thixocasting that the injection velocity and pressure may be reduced since the material to be cast is more viscous than liquid metals and fills the die more smoothly. Viscosity can also be adjusted by changing the volume fraction solidified, the primary particle size or the shearing rate.

Pressure die castings have been made successfully, with a bronze slurry (88% Cu, 10% Sn and 2% Zn) as much as 70°C below the alloy's liquidus temperature[17]. As the slurry fraction of solid level increased from 0 to 0.5, increased casting soundness was observed as a result of reduced turbulence in the die filling and reduced solidification shrinkage. Die temperatures (typically 275°C) were considerably lower than in conventional bronze die casting (350–500°C) while the casting pressures and ingate velocities required to produce good quality thixocastings were reduced.

Newton–New Haven Die Casting Company, USA, produced 380.0 aluminium alloy die castings at fraction solids above 0.4 on a 400 tonne Kux cold chamber machine[18]. Rheocasting and thixocasting techniques produced castings with surface quality and density comparable to conventional die casting. Controlling the time taken in reheating a cast slug for thixocasting, the temperature employed and the arrangement for transfer from the reheating furnace to the injection sleeve of the die casting machine were important.

Compocasting

Compocasting[23] has been exploited to produce a variety of matrix composites using high strength, high modulus discontinuous fibres to give improved mechanical properties. Particles of aluminium oxide, silicon carbide, tungsten carbide and glass in amounts up to 30% by weight ranging in size from a few micrometres to $35\,\mu\text{m}$ have been added successfully during vigorous agitation of partially solidified aluminium LM24 (A380.0) die casting alloys. Since the alloy is already partially solid, settling, floating and agglomeration is prevented and the particles are mechanically entrapped and dispersed. Composites prepared in this way can be cast when the alloy is still partially solid or after reheating. Recent work has been concerned with producing simple shapes using squeeze casting apparatus, but potential application of these composites are envisaged in the die casting process. Preliminary work at the University of Illinois showed that an aluminium alloy containing 15 wt% of Al_2O_3 particles $3\,\mu\text{m}$ in size increased the alloy's tensile strength by about 11% and wear properties by over 400%[20]. Work is continuing with increasing amounts of particle additions.

Die life and process costs

Insufficient data is available for detailed comparison of die life between semisolid and conventional die casting processes, but two significant observations have been made. First, the metal charge enters the injection sleeve as a cylinder and does not spread along the length of the sleeve as does a liquid. This reduces the contact area between metal and shot sleeve decreasing both the extent of heating and tendency for warping and distortion of the sleeve. Secondly, die thermal measurements have indicated that in thixocasting, maximum die surface heating is reduced by a factor of four, surface heating rate is reduced by a factor of seven, and initial surface temperature gradient is reduced by a factor of eight[21]. Most of the work involved bronze alloys at casting temperatures 120°C below the liquidus temperature of 1050°C. Encouraging results led to investigating die lives when thixocasting stainless steel utilizing high thermal diffusivity copper-base die materials. These had previously been unsuitable for conventional die casting of ferrous alloys because of their relatively low softening temperature. However, when thixocasting stainless steel using a copper, 0.5% chromium, 0.5% zirconium die material preheated initially at 40°C the maximum die temperature did not exceed 300°C[22]. By ejecting castings with minimum delay and reducing the heat penetration that is caused by water cooling, the need for internal cooling passages was eliminated and die cracking was reduced.

Rheocast Corporation, a commercial die casting plant at Marcus Hook, Pennsylvania is developing and commercializing the new technology for many alloy systems, the major effort being directed to aluminium alloys for valve assembly parts, electrical and decorative hardware up to 1 kg shot weight. An aluminium brake master cylinder of relatively thick section normally produced by permament mould casting was sucessfully rheocast in an aluminium alloy and subsequently heat treated. The Rheocast process is reported to represent a saving of 12% in direct manufacturing costs[23]. Melt losses with semisolid metals are considerably reduced compared to conventional die casting which aid in controlling metal composition and casting rejection losses.

References

1. BOYLE, G.R.; KAISER, W.D.
    Computer–Aided Die Design for Zinc Die casting Dies. *Society of Die Casting Engineers* Congress Paper No. G-T81-057 (1981)

2. BOOTH, S.E.; ALLSOP, D.F.
    Thermal control and design of dies. *Society of Die Casting Engineers* Congress Paper No. G-T81-056 (1981)

3. BENNETT, F.C.
    What to expect from a vacuum die casting. *The British Foundryman,* **54,** (2), 54–58, 507 (February 1961)

4. GREY, C.M.
    US Patent No. 1 153 270 (14th September 1915)

5. BRUNNER, A.J.; TREBES, B.M.
    US Patent No. 2 243 835 (3rd June 1941)

6. WALKER, H.K. (H.K. Barton)
    A survey of vacuum die casting systems. *Society of Die Casting Engineers Information Bulletin* N6 (January/October 1963)

7. HODLER, F. *(Private communication)*

8. BOMBE, K.
    Vacuum high pressure die casting of difficult high grade parts. *Information Bulletin* issued by Fondarex, Territet, Montreux, Switzerland.

9. HERRSCHAFT, D.C.; RADTKE, S.F.
    Pore-free die casting – its impact on automotive hardware. *American Society for Metals,* Report system paper No. 76–15 (October 28th 1976)

10. (Editorial)
    Elimination of gas porosity in die casting. *Die Casting Engineer,* **15,** (6), 4 (November/December 1971)

11. MIKI, I.; KIDO, T.
    Applications of pore-free die casting. *Die Casting Engineer,* **18,** (2), 22 (March/April 1974)

12. BATTISTON, L.
    Use of oxygen purge in zinc die casting. *Society of Die Casting Engineers* Congress Paper G-T77-075 (1977)

13. BATTISTON, L.
    Development of the pore-free process for hot chamber zinc die casting. *Society of Die Casting Engineers* Congress Paper No. G-T79-045 (1979)

14. RADTKE, S.F.
    Pore-free die casting – A progress report. *Die Casting Engineer,* **16,** (5), 96 (September/October 1972)

15. SPENCER, D.B.; MEHRABIAN, R.; FLEMINGS, M.C.
    Rheological behaviour of Sn-15% Pb in the crystallization range. *Metallurgical Transactions; American Society for Metals,* **3,** 1925–1932 (1972)

16. *UK Patent* Application 2 042 386 A (24 September 1980)

17. YOUNG, K.P.; BOYLAN, J.F.; BOND, B.E.; RIEK, R.G.; BYE, R.L.; FLEMINGS, M.C.
    Thixocasting copper–base alloys. *Die Casting Engineer,* **21,** 45–52 (March/April 1977)

18. (Editorial)
    Die casting semi–solid metals. *Machining and Production Engineering,* **125,** (3218), 146–150 (July 1974)

19. GIBSON, P.R.; CLEGG, A.J ; DAS, A.A.
    Compocast graphitic aluminium – silicon alloys. *Foundry Trade Journal,* **152,** (3232) (25 February 1982)

20. LEVI, C.G.; ABBASCHIAN, G.J.; MEHRABIAN, R.
    Metal Matrix Composites. Metals and Ceramics Information Center, Columbus, Ohio

21. BACKMAN, D.G.; MEHRABIAN, R.; FLEMINGS, M.C.
    Die thermal behaviour in machine casting of semi-solid high temperature alloys. American Institute of Mechanical Engineers, 105th Annual Meeting, Las Vegas (23 February 1976)

22. YOUNG, K.P.
    Extending Die Life *Proceedings of the Workshop on Rheocasting,* pp. 79–84. Metals and Ceramics Information Center, Colombus, Ohio

23. HESS, W.T.
    Rheocast Corporation a Commercial Venture. *Proceedings of the Workshop on Rheocasting,* pp. 121–123. Metals and Ceramics Information Center, Colombus, Ohio

Chapter 25

Pressure die casting dies

A pressure die casting die is an assembly of materials, mostly ferrous metals, each of which plays a part in a mechanism which will operate under conditions of rapidly changing temperatures, as the molten metal is injected under pressure and then immediately cooled. Some parts of the die merely act as holding elements; others, such as the die inserts and the cores, have to withstand the impact and high temperature of the molten metal. The mechanisms for moving the ejectors and cores must work smoothly in the constantly changing temperature of a die. Parts which act as bearings are made either of non-ferrous metals such as phosphor bronze, or the steel surfaces are given a nitride, Tufftride or other treatment to resist wear.

The die insert for producing the shape of the casting has to be machined accurately and heat treated to provide the best possible mechanical properties at high temperature. Most die casting dies operate automatically and, while this leads to high productivity when there are no delays, the cost of breakdowns are correspondingly high. During the past decade the manufacture of dies has been revolutionized by sophisticated methods ranging from improved spark erosion to computer-controlled die making. Among other efforts the Science Research Council[1] is supporting a large scale research programme to include die heat treatment and surface coatings, computer-aided die design, special machining processes and the economics of die manufacture.

A single die may contain 10 or more different steels plus several non-ferrous metals and special heat resisting alloys. *Figures 25.1* and *25.2* illustrate schematically a die which typifies the various components, each of which involves design, engineering and metallurgical problems.

Low alloy steel and cast iron components

The ejector box is constructed of several rectangular blocks built up to contain the ejector plates which are guided by round section runner bars or guide pillars. The box sections, ejector plates and runner bars are of mild steel, with approximately 0.15% carbon. The guide pillars and ejector stops may be case hardened. The function of these parts is to support and guide, not to endure shock loading. Bolsters undergo mechanical impact and stress but not a great deal of thermal shock and are often made of medium

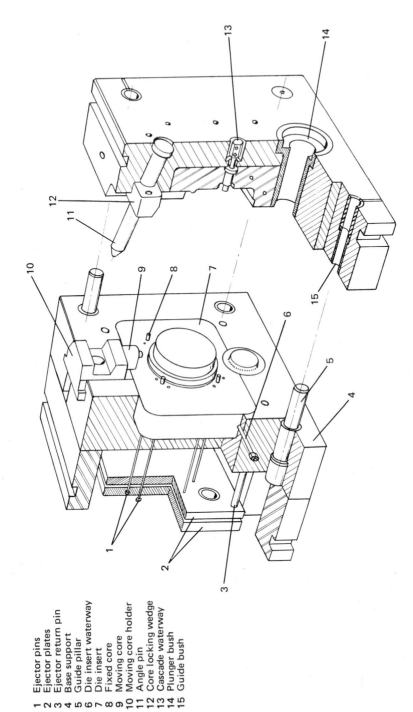

1 Ejector pins
2 Ejector plates
3 Ejector return pin
4 Base support
5 Guide pillar
6 Die insert waterway
7 Die insert
8 Fixed core
9 Moving core
10 Moving core holder
11 Angle pin
12 Core locking wedge
13 Cascade waterway
14 Plunger bush
15 Guide bush

Figure 25.1 Typical die assembly

249

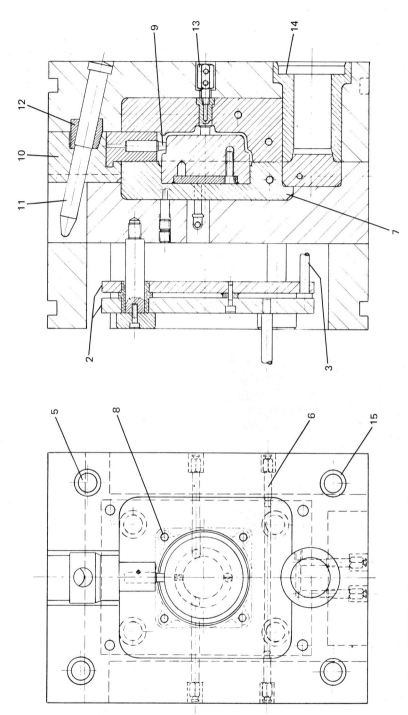

Figure 25.2 Views illustrating structural features of die shown in Figure 25.1

carbon steel. Alternatively these parts are steel castings or they may be of a spheroidal graphitic cast iron.

The die blocks contain the movement mechanisms of the die and the inserts incorporating the casting form so that the assembly can be mounted between the tie bars of the machine. The blocks are usually made of medium carbon steel; a typical British specification is BS 970 080M40 (En8), with 0.4% carbon, 0.8% manganese and 0.3% silicon. Sometimes a prehardened steel is used, a typical analysis being 0.35% carbon, 1.0% manganese, 0.5% silicon, 1.65% chromium and 0.5% molybdenum. This composition is covered by the American AISI specification P.20.

The die blocks are shaped to form the recess into which the inserts will be fitted. Alternatively, and this is a procedure which is steadily gaining on the other method, large die blocks are supplied as medium carbon steel or cast iron, cast as closely as possible to the required shape. The blocks are ground flat so that the injected alloy will not escape under the considerable pressure of injection. It is preferable to have the parting line in one plane, but sometimes the design of a component makes it necessary to construct the die with irregular parting line surfaces to avoid creating an undercut. Once the parting line has been established and the blocks machined, guide pillars, usually between 10 mm and 60 mm in diameter, depending on the size of the die blocks, are incorporated into one half of the die to ensure exact alignment of the two die halves. Holes of an appropriate size are machined into the other die half and fitted with bushes, normally of carburized steel. On symmetrical dies, one pillar may be offset, to avoid assembling the die incorrectly. The pillars are normally of case hardened mild steel; occasionally case hardened nickel steels are used, for increased strength.

Where possible, guide pillars are fitted into the die half which holds any protruding die form (usually the moving, or ejector, half) to give protection to the cavity form when the die is taken from the machine for maintenance. If positioning in this way causes interference with a casting retrieval arm, particularly if it is of the fixed path extraction type, unable to offer lateral movement, one pillar is repositioned in the fixed half of the die, as shown in *Figure 25.1*. Square guide pillars are often used with larger dies to be operated on machines of 600 tonnes locking force and upwards. This system makes for easier adjustment, sometimes required on such dies, arising from thermal expansion of die components.

The die insert

This is the heart of the die, since it contains the outside shape of the component, forming the cavity into which the molten metal will be injected. Die inserts, together with the cores which form recesses, must endure the effects of temperature and injection pressure of the molten metal. Each injection is a step towards the thermal fatigue which will eventually cause deterioration of the die. The cycle of stresses in the die insert results in a sudden increase of temperature as the molten metal enters, followed by a rapid decrease. Often the die insert is built up from several pieces, either to save material when large blocks incorporate small projections, or to facilitate replacement of parts, or to ensure that an available heat treatment furnace will accommodate the die block. Spark erosion techniques, widely used nowadays to manufacture small or medium sized die components in one piece, avoid the danger of molten metal being forced down joins between separate parts.

Cast-to-form[2] is now a well established procedure for making cast iron permanent mould dies, of sufficient accuracy for the requirements of that process. Die inserts for pressure die casting need to be made to an accuracy of the order of ± 0.002 mm/mm. Furthermore, the inserts are usually to be made of 5% chromium steel and, so far, the difficulties of casting such a material to a sufficient accuracy has prevented 'cast–to–form' from being used for pressure die casting dies.

Where deep inserts are required, clearances to assist fitting are provided and radii on the corners are included to improve strength and remove stress raisers, as indicated in *Figure 25.3*. For very deep inserts, filleting at two levels is advisable, both for ease of

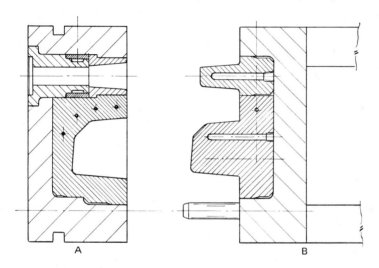

Figure 25.3 Example of insert fitting. (a) Very deep cavity fitting at two levels. (b) Shallow cavity

assembly and to assist heat transmission from the insert to the die block. The selection of steel and its subsequent heat treatment are vital factors in ensuring optimum life and accuracy of the die casting die.

Since aluminium represents the major part of the industry's output, it is natural that consideration of die steel problems centres around the production of aluminium alloy die castings. The essential requirement of a 'hot die steel' is to provide maximum endurance under conditions where molten metal is injected into the die, causing rapid fluctuations of temperature. These cause the die insert to expand and contract against the constraint of the die blocks which contain it. Ideally the die steel should have a thermal expansion as low as possible (to minimize the expansion and contraction), as high a modulus of elasticity as possible (to minimize the stress caused by the expansion and contraction) and a maximum fatigue strength at elevated temperatures (to delay the commencement of craze cracking). Other factors such as the machinability of the steel and its cost are important, but they are only relevant when the selected steel gives long die life.

The following *Table 25.1* lists seven hot die steels which are currently in use. In the UK they are included under the specification BS 4659 and given the BH prefix. In the USA similar steels come under the AISI series of specifications and are given an H prefix. There are other series of specifications in other countries, but the 'H' or 'BH' prefixes are known internationally. There has been an increasing tendency to concentrate on the use of one die steel, and H. 13 is favoured. Such standardization has advantages in purchase policy and stock control. The steels are marketed with a variety of trade names.

TABLE 25.1

Specification	Constituents (%)							
	C	Si	Mn	Cr	Mo	V	Co	W
BH10	0.35	1.0	0.4	3.0	2.8	0.5	–	–
BH10a	0.35	1.0	0.4	3.0	2.8	0.75	3.0	–
BH11	0.4	1.0	0.4	5.0	1.5	0.5	–	–
BH12	0.35	1.0	0.4	5.0	1.5	0.5	–	1.5
BH13	0.4	1.0	0.4	5.0	1.5	1.0	–	–
BH19	0.4	0.4	0.4	4.25	0.4	2.2	4.25	4.25
BY21	0.3	0.4	0.4	2.75	0.6	0.4	–	9.3

Most of these steels contain alloying elements chromium, molybdenum and vanadium, their primary function being to enhance the properties through control of the microstructure in conjunction with suitable heat treatments. As discussed on page 276, alloying elements confer additional hardenability and this effect is their most important function. Small amounts of several alloying elements are more effective in increasing hardenability than a relatively large amount of a single element.

Chromium, molybdenum and vanadium are all carbide-forming elements and they prevent softening of the steel on tempering or stress relieving. Chromium and molybdenum carbide particles give good wear properties to the steel and increase its strength. Vanadium forms a strong high melting point carbide which acts as a grain refiner, preventing grain growth during heat treatment.

In the 1960s tungsten steels were favoured since it was thought that, as tungsten provided long life for cutting tools, this element should also give long life to a die casting die steel. Tungsten steels perform well at constant elevated temperatures but in severe temperature cycling situations, such as pressure die casting, they have a low thermal shock resistance and failed prematurely. Their use declined in favour of alloy steels such as H. 13.

Die inserts for zinc alloys

Until the development of automation, dies for zinc alloy die castings were often made with mild or medium carbon steels. Then die inserts were made of chrome–vanadium steel, a typical composition being 0.5% carbon, 0.65% manganese, 1.1% chromium and 0.2% vanadium, covered by the American AISI specification L2. Before automation, chrome–vanadium steel die inserts, operating at a hardness of only 32 HRC, were adequate; they could be supplied in a pre-hardened condition but were soft enough for machining. The development of automation from the 1950s onwards emphasized the need for steel endurance under the conditions of rapid production and the importance that no die wear occurred to cause metal to adhere to cracks or joints in the die, leading to delayed production. It was then found preferable to make die inserts for zinc alloy in the same 5% chromium steel used for aluminium alloys, to H.13 specification and hardened to between 45 and 48 HRC.

Electroformed die inserts

In recent years there has been some interest in the manufacture of dies using electroforming techniques; new developments have extended their use into zinc die casting[3]. A mandrel having the same form as the die component to be made is produced in epoxy resin using a model or pattern as the master and its surface is metallized with a layer of silver about 1μm thick. The mandrel is then placed on a jig and immersed in a nickel sulphamate plating solution to which cobalt sulphamate has been added. Electrodeposition continues until the electroform is 3—5 mm thick. It is then separated from the mandrel after the edges have been machined to remove unwanted build up. The back of the electroform die insert pieces are irregular in shape and they must be embedded in a backing which can be machined to a regular shape and then fitted in the die block. One method of doing this has been to electrodeposit copper on to the back of the electroformed die face to a thickness of about 50 mm but this takes several weeks. Recently, advances[4] have been made in backing electroformed moulds by metal spraying aluminium— silicon alloy, copper or bronze. This process takes only a few hours and produces a thick deposit with good mechanical properties.

Surface hardness levels of 30 HRC are typical but can be increased by the addition of organic agents to the electrolyte. Many such agents incorporate sulphur, which embrittles the deposit at temperatures in excess of 200°C, while a nickel alloy deposit containing up to 15% cobalt produces hardness of 44 HRC without the need for sulphur-containing agents. The production of electroformed die insert has been the subject of much research during the past 10 years. Considerable success has been achieved[5] in the plastics industry, type metal moulds, bank note printing plates, press tools and several branches of food processing. Experimental production has been done in the manufacture of zinc die casting die inserts, but so far the success has been only mildly encouraging.

Dies for brass die castings

When the die casting of brass entered into the repertoire of the industry, higher die surface temperatures were encountered than with aluminium and it was found that 5% chromium steels gave only limited die lives. Alloy steels containing chromium, molybdenum and vanadium, sometimes with cobalt as in BH10a, sometimes with tungsten, as in BH21 and sometimes with cobalt and tungsten, as in BH19 were used. In more recent years die inserts for brass have been made in special heat resisting materials, described on page 158.

The cores

Holes and recesses in the casting are produced by cores, usually mounted in the moving die half. Like the die inserts, they require properties which enable them to continue operating smoothly and accurately for a maximum number of shots. However, in contrast to the die insert, the replacement of cores that have become worn is not so costly, and dies that are required to operate on long runs are provided with ample spare cores especially those of small diameter. Cores are made of alloy steel, heat treated to provide maximum endurance. Where core applications call for materials with improved ductility and toughness, the 18% nickel maraging steels are specified (*see* page 255). In brass die casting, there is a growing interest in the use of cores in molybdenum and tungsten-based refractory metals which are capable of conducting away large amounts of heat.

There are several methods for assembling cores within a die, whether the cores are fixed or moving. For our model die in *Figure 25.2,* the short fixed core is secured by means of a grub screw at the rear head of the core and a second grub-screw, used for locking purposes. This arrangement enables quick core removal and replacement if damage occurs. Another method of securing short protruding cores where replacements are not often envisaged is to recess the core head position into the holding plate which is held in position by the associated die block. Flats on core heads or securing pins and keyways are often employed to prevent the cores from turning. Where several cores are required close together there may not be sufficient space for individual holding arrangements and it is common practice to secure the group of cores by means of a retaining plate.

Fixed cores are used when the axis of the cored shape is at right angles to the parting line of the die. Moving cores are required when the axis of the cored recesses are at any other angle within the die parting line. Referring to the model illustration, a moving core is actuated by means of an angled dowel pin, normally made from toughened and nitrided steel as detailed in *Figure 25.4.* The core is supported within a core slide, often made from chromium—vanadium steel toughened and nitrided to prevent drag or eventual seizure. Guideways for toughened slides are machined within the bolster block.

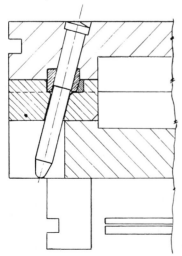

Figure 25.4 Core movement employing angled dowel pin

Angled dowels are often used to withdraw moving cores on small dies but, in larger dies, the use of hydraulic cylinders is a common method. When several moving cores are incorporated within a die it is often preferred to use hydraulic cylinders since access between the die space is improved, enabling easier casting retrieval. When short core movements are involved, a toggle mechanism can be operated by a hydraulic cylinder. Some casting configurations require complex core movements that may be rotary or spiral and these can be accommodated by special purpose mechanisms.

Maraging steels

When severe conditions require greater heat resistance and thermal fatigue enduring properties than can be provided by alloy steels such as H.13, special materials are necessary; among these the maraging steels are becoming widely used[6]. They offer a combination of high strength and toughness, making them suitable for cores in aluminium alloy die casting. The name is derived from a combination of the words 'martensite' and 'ageing'. They are low carbon steels produced by vacuum melting from high quality materials and contain nickel, molybdenum, cobalt, plus smaller but important additions of titanium, aluminium and silicon. Four compositions are in general service and are shown in *Table 25.2*, classifed in terms of proof stress, ranging from 1400 to 2400 N/mm². The alloys are available from several suppliers under their own designations.

TABLE 25.2

Grade	C	Mn	Si	Mo	Ti	Al	Ni	Co
18 Ni 1400	0.03	0.1	0.1	3.0-3.5	0.15-0.25	0.05-0.15	17-18	8-9
18 Ni 1700	0.03	0.1	0.1	4.6-5.1	0.3-0.5	0.05-0.15	17-19	7-8.5
18 Ni 1900	0.03	0.1	0.1	4.6-5.2	0.5-0.8	0.05-0.15	18-19	8-9.5
18 Ni 2400	0.01	0.1	0.1	3.5-4.0	1.6-2.0	0.1-0.2	17-18	12-13

(Manganese and silicon are in effect 0.12% maximum but the total Mn plus Si must not exceed 0.2%)

Annealed maraging steels[7] are relatively easy to machine and in that condition they have a hardness of about 30 HRC. After initial machining, solution treatment is applied by heating to 800/900°C and cooling in air or oil at a rate fast enough to prevent age hardening. After final machining the age hardening treatment for all grades is performed by heating to 480°C. Depending on the section thickness of the treated part and the final hardness required it is held at this temperature for 6-9 h, followed by air cooling. Hardness levels achieved are about 50-54 HRC, while holding for shorter times at higher temperatures will give a lower hardness.

The main hardening precipitates include compounds of nickel—titanium and iron—molybdenum with small amounts of nickel—aluminium. During the age hardening process, maraging steels shrink; these dimensional changes following treatment at 480°C are shown to be less than 0.0004 mm/mm for grade 1400 and less than 0.0008 mm/mm for grade 1900. The actual figures indicate a dependence on titanium content. Higher ageing temperatures will lead to higher levels of shrinkage, which must be allowed for in machining. Maraging steels used for cores and small die inserts have given good performance but abnormal wear has occurred on large inserts used in aluminium die casting, due to high die surface temperatures and the erosion effect of aluminium. Maraging steels are used satisfactorily for weld repairs on die steels. Unlike welding with H.13, no die preheating is necessary, in fact a cold die is necessary to cool the maraging weld rapidly. This is then deposited by successive small amounts as a soft and ductile material to decrease the chances of weld cracks forming. Low temperature age hardening of the weld following machining also serves to temper the prehardened zone which is formed in a hot die steel adjacent to the weld. General properties of the parent die steel are not affected by welding.

The cooling passages

Certain sections of the die, particularly at the metal entry areas and those adjacent to heavier casting sections, must be maintained at a suitable temperature to ensure optimum casting production rates and die life. Slides, cores and other moving components of the die must also be cooled to avoid drag or seizure and ensure smooth operation. Water is commonly used as the cooling medium, although there have been some advances in the use of oil for both die preheating and subsequent die cooling during casting (*see* page 215) and the use of heat pipes. Cooling passages are made by drilling a series of holes into the main insert blocks either all the way through or, as was shown in *Figure 25.2*, by inserting plugs in certain portions of the holes. Water then circulates through the channels before flowing out through an exit pipe. The position and size of cooling channels have a marked effect on the life and performance of the die. If weakening of the insert block is to be avoided and adequate temperature control achieved, cooling passage centres should not be closer than 25 mm from the cavity face. To avoid the risk of subsequent die cracking, cooling passages must have a smooth surface finish free from rough machining marks and be so arranged that there is no leakage, particularly at parting line surfaces of the die. Use of hard untreated water should be avoided, since even small deposits of scale will lower the heat extraction rates and involve frequent descaling. Similarly, no salt residues should be retained in cooling channels following hardening operations. A solenoid valve can be employed to control water flow automatically, monitored with the die casting machine cycle.

Often, water cooling needs to have access to localized die areas including long cores which are often 'hot spot' areas. Typical water passage systems for these applications, termed 'fountains' or 'cascades', are shown in our model die on page 248 and, in more detail, in *Figure 25.5*. A flow of water is directed locally behind a core or die area and its

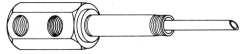

Figure 25.5 Water cooling cascade

return is confined to an outer sleeve, the whole unit taking up little space and being conveniently threaded for use of assembly. Alternatively, heat pipes[8] can be used to cool 'hot spot' areas within a die; their use has increased in recent years and they are available in standard sizes from a number of manufacturers. Their heat extraction capability, several hundred times greater than that of copper, make them particularly suitable for extracting large amounts of heat leading to increased production rates and improved casting quality. *Figure 25.6* illustrates schematically the operation of a heat pipe. A sealed tube, normally

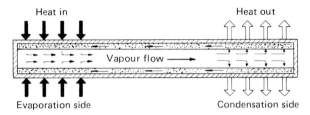

Figure 25.6 Schematic diagram showing the mode of heat transfer within a heat pipe during operation

of copper and externally plated with nickel or tin for corrosion protection, contains a wick of absorbent material lining the inside wall in which a capillary action can be generated. Several heat transfer working fluids can be used, although water is found to be adequate for zinc and aluminium die casting. Fluids capable of working at higher temperatures are available for brass and ferrous die casting. As heat is applied to the end of the heat pipe, the liquid in the wick evaporates; vapour flows to the cooler regions of the heat pipe where condensation occurs and the cooled liquid travels back to the evaporator section for continuous circulation. Sizes of pipe range from about 2–12 mm diameter and 50–250 mm lengths; the larger sizes may be used in the main body of the die to even out temperature gradients, whilst the smaller sizes can be used to reach into deep narrow recesses where access to water 'cascades' is restricted. Surface scale does not develop in the sealed pipe and its cooling efficiency is maintained. Heat can be dissipated from the condenser section of the heat pipe by air or adjacent die components while an alternative method is to position the condenser section of the pipe in circulating water channels.

Ejectors

Within a few seconds of the metal being injected into the die cavity, the casting is already solidified and cooling rapidly; the moving die half is retracted, thus leaving an opening between the die halves. The ejectors are then moved forward as part of a timed production sequence and push the casting from the moving die half. This forms the typical round ejector pin marks characteristic of pressure die casting. The steel normally used for ejectors is a 5% chromium (H.13 type), heat treated to give a hardness of between 38 and 42 HRC. Some ejector pins are supplied in a nitrided condition and some are treated by Tufftriding as described in Chapter 28. Nowadays ejector pins, shown along with other standard die components such as bolster plates, dowels and bushes, are hardly ever made by the die casting companies since they can be purchased ready made more cheaply and with more consistent quality from outside specialists.

It is essential that the ejectors are machined precisely and that the ejector guides in the moving die half are accurately ground with the necessary clearance between guide and pin.

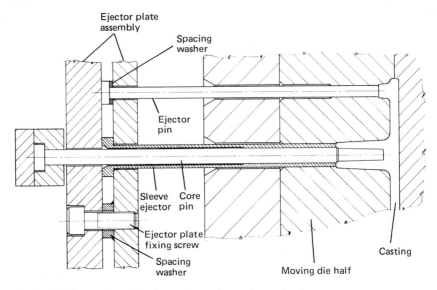

Figure 25.7 Arrangement for ejector pins and core sleeve ejection

The ejectors are disposed in positions so that they will press evenly against the casting without distorting it; indeed, selection of the best positions for ejection calls for long experience and the ability to visualize the stresses which will be exerted on the cooling die casting. Quite often, casting overflows or special ejector pads which are later removed, provide the necessary surface on which ejectors can push the casting clear from the die, so avoiding ejector marks on the casting and offering additional ejection around the periphery of the cast part. When castings need to be ejected from long cores or when their shape or small size make it impractical to use ejector pins or ejector pads, it may be necessary to employ a sleeve ejector, the most common being a tube around the fixed core pin which ejects the casting from the core after die opening. These have been used as part of the development in 'cast-to-size' die castings, where improved accuracy and the elimination of machining processes have been achieved. Whichever ejection system is used it is important that the pin tolerances and alignment with the various parts in the die are accurate. Automatic die casting, with its increased speed of production, relies on trouble free operation if maximum utilization is to be achieved; if ejection systems are badly made and located, they can be a source of continual irritating production delays.

Refractory alloys

There has been a gradual increase in the use of high melting point refractory alloys of tungsten[9] and molybdenum[10]. Developed for specialized applications, they are more costly than hot work steels and would not be recommended where a steel part now functions satisfactorily. The main attraction for their use arises since their thermal conductivities are several times higher than steel while thermal expansion is about half. The combined effect of these properties leads to good resistance to thermal fatigue, soldering and erosion and their applications range from cores in aluminium die casting dies to complete inserts in dies for the higher melting point brass and ferrous casting alloys (*see* pages 158 and 162). The high thermal conductivity of these materials ensures that for most core applications water cooling is not required.

Before embarking on the use of refractory alloys it is important that die designers and toolmakers be well advised on their properties and correct use. Due to the differences in thermal expansion between a refractory alloy part and its adjacent steel holder or ejector pin mechanism, clearances need to be adjusted to ensure smooth die operation. To achieve the full benefits of high heat transfer rates the back face of any insert or core and its matching face of the die steel holder must be clean and smooth to ensure firm contact. Refractory alloys show more notch sensitivity than die steel and to reduce any possibility of fracture, sharp inside corners and notches should be avoided and generous radii provided wherever possible.

Tungsten alloys

A tungsten-base alloy developed by the Mallory Company in America during the 1960s is known as Anviloy. In Britain it is marketed by Johnson Matthey Metals Ltd., and known as Matthey 4000. The nominal composition of this and a similar alloy are shown in *Table 25.3*.

TABLE 25.3

	Matthey 4000	Matthey 4100
Tungsten	90.0	86.0
Nickel	4.0	6.7
Iron	2.0	3.3
Molybdenum	4.0	4.0

The lower tungsten content of the 4100 alloy confers additional ductility. Belonging to a family of materials generally referred to as 'tungsten-base high density composites', the high melting points of tungsten and molybdenum call for production by powder metallurgy techniques to obtain alloy density levels greater than 90%. Intimate mixtures of fine particle size powders are compacted using either isostatic or closed die compaction techniques. A sintering treatment at about 1500°C results in densification of the structure due to the interaction between the nickel–iron liquid phase and the refractory metals tungsten–molybdenum by means of a solution-precipitation-growth mechanism. The resultant uniform structure of this high density composite consists of tungsten grains surrounded by the nickel–iron–molybdenum matrix. No heat treatments are necessary for these alloys. They will conduct heat about four times faster than steel and their thermal expansion is less than half that of steel. At 650°C, the tensile strength of the alloys is comparable with H.13 die steel, but at more elevated temperatures they are stronger. Machining can be carried out using tungsten carbide-tipped tools, but since tungsten is removed only slowly by spark machining methods and with high electrode wear, conventional machining is employed as much as possible. Welding can be done by the inert gas process using tungsten alloy welding rods, and employing a protective atmosphere until the material has cooled below red heat. In a paper by K.W. Simms[11] details of alloy properties, applications and cost factors are presented.

Molybdenum alloys

The composition of two alloys containing over 99% molybdenum developed by the Climax Molybdenum Company are shown in *Table 25.4,* the alloy Mo 0.5 Ti has a slightly lower strength than TZM.

TABLE 25.4

	TZM	*Mo 0.5 Ti*
Titanium	0.50	0.50
Carbon	0.015	0.020
Zirconium	0.08	—
Molybdenum	Balance	Balance

Production of these alloys involves compaction of metal powder, sintering and casting under vacuum; the absence of any phase changes makes heat treatment unnecessary. Their hardness and strength is achieved by a strain hardening or cold working mechanism, a solid solution hardening effect from the action of alloying elements dissolving in the molybdenum, and dispersion hardening due to precipitation of complex carbides, leading to strength at high temperatures. Both alloys have thermal coefficients of expansion about half that of H.13. Their thermal conductivity is several times higher than that of H.13, but varies considerably according to temperature. At about 650°C, their conductivity is about eight times that of steel, while at 1000°C the conductivity is about four times.

Their tensile strengths and hardness are superior to H.13 at 650°C. Machining is by techniques similar to those employed for hot work die steels. Although acceptable welds have been achieved with sheet structures utilizing inert gas, ultrasonic and electron-arc beam techniques, welding of die casting die inserts in these alloys should be reduced or avoided. A paper by R.W. Burman[12] discusses the properties of molybdenum alloys and their application in die casting.

Nimonics

These alloys were developed during the Second World War for applications in jet aircraft[13]. They are based on the electric resistance wire composition 80% nickel, 20% chromium, to which other elements are added. Nimonic 75 contains about 0.4% titanium and 0.1% carbon. In Nimonic 80a and some other alloys in the series, precipitation—hardening takes place by a complex nickel—aluminium—titanium phase. Nimonic 90 is a nickel—chromium—cobalt alloy to which small additions of titanium and aluminium are made, and there are others in the series suitable for very high temperature service. Nimonics 75 and 80a are used in permanent mould casting of aluminium bronze for cores or die inserts. In view of this success it is disappointing that so far all tests with Nimonics in pressure die casting dies have not shown worthwhile results, largely due to grain boundary attack.

References

1.   (Editorial)
     Modern methods adopted in die and mould manufacture. *Metals and materials,* 38–41 (January 1977)

2.   YOUNG, W.
     Die building and maintenance. *Foundry,* **96,** (10), 180 (October 1968)

3.   WEARMOUTH, W.R.
     Application of new developments in electroforming technology in the toolmaking industry. *Plastics and Rubber Processing,* **2,** (4), 131–138 (December 1977)

4.   DEAN, A.V.
     Further developments in the production and use of cast and sprayed backings on electroformed moulds and dies. *Metallurgia,* **45,** (5), 243–248 (May 1978)

5.   WATSON, S.W.
     Applications of nickel electroforming in Europe. *Electroplating and Metal Finishing,* **28,** (7), 3–11 (July/August 1975)

6.   KRON, E.C.
     Maraging Die Steels. *American Die Casting Institute,* Des Plaines, Illinois (1968, Spring technical seminar)

7.   (Editorial)
     Heat-treatment procedures. 18 – 300 maraging steel for diecasting. *Die Casting Research Foundation, Technical Bulletin* 01-74-010 (1974)

8.   REAY, D.A.
     Heat pipes, a new diecasting aid. *Foundry Trade Journal,* **143,** (3125), 1161–1166 (24 November 1977)

9.   FRAZIER, R.T.; BORBELY, A.; LAWRENCE, J.R.
     Tungsten–base materials for die casting. *Society of Die Casting Engineers* Congress Paper No. 0372 (1972)

10. BURMAN, R.W.
     Performance of new molybdenum–base alloy tools in casting processes. *Modern Castings,* **46,** 471 (August 1964)

11. SIMS, K.W.
     Design and application of refractory alloy diecasting tools. *Foundry Trade Journal,* **132,** (2890), 597–606 (27 April 1972)

12. BURMAN, R.W.
     New die casting cores show long life. *Precision Metal Molding,* (June 1962)

13. (Technical brochure)
     Nimonic alloys. Henry Wiggin Co. Ltd., Hereford. Publication 25048 E (March 1973)

Chapter 26

Thermal fatigue of die casting dies

Dies for the alloys of lead and of zinc usually continue to give trouble free service for more than a million shots. Aluminium and magnesium alloys, cast at about 200°C higher than the melting point of zinc, give die lives of about 50 000 shots — sometimes much more, occasionally substantially less. When brass is die cast, using heat resisting materials, over 10 000 shots are expected, although die lives several times that figure have been quoted, such as the example mentioned on page 158. When stainless steel is die cast, using molybdenum die inserts, about 5000 shots are achieved before resinking is necessary. Thus the melting points of the alloys which are die cast have a considerable influence on the die lives which are obtained.

By far the greatest number of die castings are made in aluminium alloys, and one of the most aggravating problems in their production is the deterioration of dies. Having taken many hundred hours of skilled toolmakers' work, the use of sophisticated die making machines, careful heat treatment and meticulous inspection, a new die is put into production. Usually it remains in good condition for over 20 000 shots but, sooner or later, the die surface begins to deteriorate and 'hair line' or 'craze cracking' develops. By the time a typical die has produced 50 000 aluminium alloy die castings there is considerable deterioration, and extra finishing operations are needed if any part of the component requires a good surface appearance. Fortunately, many large aluminium components for engines or transmissions are not visible in the automobile, so die cracks reproduced on the casting are not detrimental. However a die which has surface defects operates less efficiently than an unblemished die, particularly in automatic production.

For these reasons, the need to study the causes of die deterioration has become more pressing than ever before. Improved die steels and careful control of their quality, checked to ensure that the die blocks are free from cracks or inclusions, have led to an improved starting point. New and improved heat treatment processes, surface treatments, greater care in the finishing of dies and more precise control of the temperatures in the casting cycle have helped towards better performance of die casting dies. Nevertheless, it is apparent from various studies that dies are not achieving the length of service which are indicated by theoretical considerations.

Usually the deterioration of a die insert or a large core proceeds by the gradual development of 'craze cracking'. Sometimes deterioration comes suddenly; such catastrophes are caused by faults in the die steel, carelessness in the manufacture of the die or incorrect heat treatment. Recently, the BNF Metals Technology Centre examined a number of dies which had failed prematurely. The following four 'case histories' are taken from their report[1] and exemplify some causes of die failure. All the illustrations (Figures 26.1–26.11) are shown by courtesy of the BNF Metals Technology Centre. *Figure 26.1* and *26.2* illustrate a pair of die inserts which failed by downright carelessness. The inserts

Figure 26.1 Pair of die inserts which failed due to lack of heat treatment

Figure 26.2 Detail of die insert showing heat checking

were used for only 15 000 shots but heat checking had already appeared after only 3000 shots. The inserts had suffered some erosion, and heat checking was most severe adjacent to the gate area as shown on *Figure 26.2*. Sections were cut from two of the inserts and Vickers hardness tests taken with a 20 kg load. The hardness range was 213–236 HV20, corresponding with a Rockwell C hardness of 13.8–19.4. The sections revealed that the structure was of an annealed steel and it was, therefore, evident that the degradation of the inserts was because samples had been taken from the die before heat treatment and then the die had been put into service untreated. Any die casting management which has been subjected to customer pressure knows that they have to be very positive in resisting demands to test dies before heat treatment and to produce castings urgently, but an example like this shows the results which may follow.

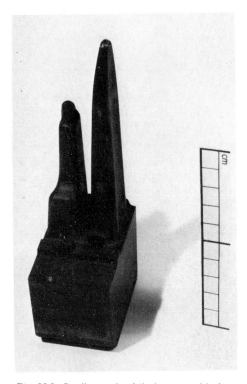

Fig. 26.3 Small core that failed, reassembled

Figures 26.3 and *26.4* illustrate the failure of a small core. The fractured surfaces consisted of three sections, a coarse grained area containing inclusions, a fine area and a medium grained area forming the bulk of the fractured surface. The failure appeared to have originated in the coarse layer of inclusions shown in *Figure 26.4* (the core is shown reassembled in *Figure 26.3*). Examination of the specimens showed that a weld repair had caused a coarse-grained structure, shown in *Figure 26.5* and *26.6*. The welded section was about 1.5 mm deep The steel adjoining the weld was hardened to 52 HRC and the bulk over-tempered to a hardness of 40 HRC. The failure of the core was therefore due to a defective weld at the base of the core; there were two inclusions, each about

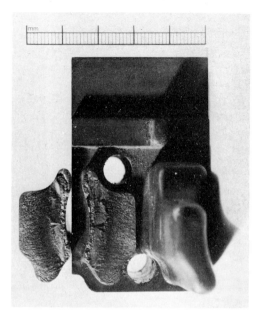

Figure 26.4 Parts of failed small core, showing inclusions

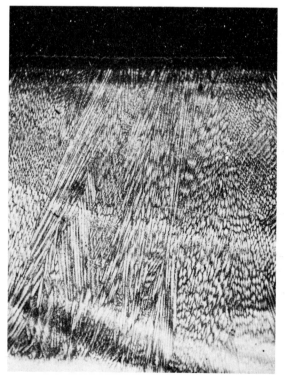

Figure 26.5 Microstructure of the small core (Figure 26.4) weld metal (×50)

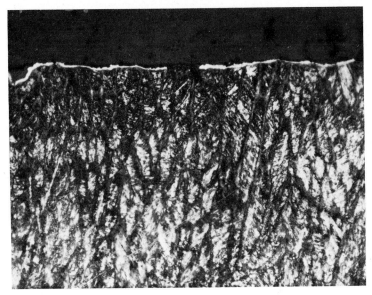

Figure 26.6 Weld metal surface of failed small core

0.5 mm diameter, situated just below the surface of the weld metal. To quote the BNF report, 'repair welding of die components is not desirable but is often necessary. When welding is carried out great care must be taken to preheat before welding and to stress relieve afterwards, otherwise cracking is likely to occur even before the die is back in service.'

Figures 26.7 and *26.8* illustrate a die insert; the point of origin of the failure lay on the radius of the island separating two runners approximately 10 mm from the edge of the shot hole. The crack then spread to a point halfway down the section shown in *Figure 26.8* and along the back of the insert to the lower part of the same illustration. Final catastrophic failure occurred when the crack extended from that position to the die face. Hardness tests were made of four sections of the die using a Vickers test with a 50 kg load which gave a hardness range of 493–513 HV50 (equivalent to 48.5–49.7 HRC).

Figure 26.9 shows a section through the area adjacent to the point where failure was initiated and with a crack about 0.5 mm deep parallel to the fracture surface. The microstructure of the steel showed tempered martensite with grain boundaries, highlighted by relief. There were small spherical particles of carbide within the grain boundaries. The failure was due to a quench crack situated on the radius of an island approximately 10 mm along the side of a runner. The hardness of the die gave an average of 49 HRC, which is greater than the hardness of 46 HRC normally recommended for this size of die and, therefore, the toughness of the steel was reduced. The failure of the die under these conditions was aggravated by the large size of the insert, 600 mm long, which caused heat treatment problems and exaggerated the thermal gradients and distortion during casting, so producing high stresses the quench cracks which initiated failure.

The fourth example is illustrated on *Figure 26.10*. It was a small die insert; the crack, indicated across the raised section of the block, appears to have spread from a sharp change in section at the base of the raised area towards the back of the block and probably originated in this sharp corner either as a single crack or as the amalgamation of

Figure 26.7 Large die insert which failed by cracking

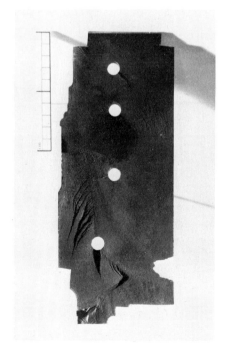

Figure 26.8 Cracking in die insert

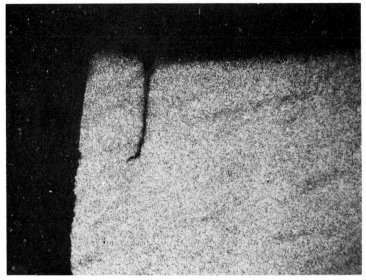

Figure 26.9 Failed die insert section adjacent to fracture surface

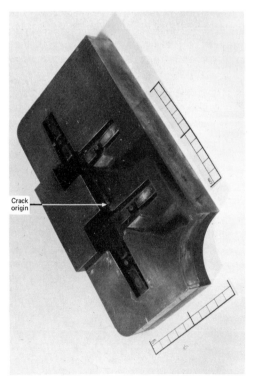

Figure 26.10 Small die insert which failed due to hardness being too high

Figure 26.11 Microstructure of steel insert (Figure 26.10) (×600)

a series of small cracks. The hardness of the block, measured using Vickers hardness tests with a 50 kg load, was in the range 595–626 HV50 (equivalent to 55–56 HRC). The structure of the steel, shown in *Figure 26.11,* was of sharply defined martensite with retained austenite and fine spherical particles, symptomatic of the untempered condition of the steel, indicated by the high hardness, which led to lower toughness aggravated by the sharply machined fillet radius which was a stress raiser. It was diagnosed that either the steel had not been tempered at all or it was tempered at too low a temperature. Initiation of the cracking either started at a quench crack following hardening of the die (although no quench cracks were detected) or as a result of stresses imposed during service. Certainly the performance of this die insert would have been much better had the tempering been done correctly, bringing the hardness to a suitable figure of 46 HRC, and if the stress raiser had been removed.

All of the examples described illustrate die failures which could have been prevented by greater care, but failures occur occasionally even in the best regulated die casting establishments. However, even if heat treatment is done correctly, repair welding super-vised carefully and if die construction is monitored to eliminate stress raisers, the life that is obtained from dies in aluminium die casting is still depressingly short.

Three causes contribute to die failure. The abrasive action of the metal injected into the die at high velocity causes erosion, and chemical attack of the die surface occurs as the 'soldering', which takes place if part of the cast metal adheres to the die. However the most frequent cause of die failure arises from thermal fatigue.

Fatigue failure

A metal subjected to a repetitive or fluctuating applied load will fail at a stress lower than that exerted by a static load. The fatigue strength of steels decreases with increasing temperature and failure can occur suddenly after a period of service. As a result of the divergence in temperature between the surface and interior of a die during casting, localized thermal stresses are developed as the expansion of the surface skin is restrained by the mass of the die. Continued thermal cycling causes cracking to develop.

Fatigue failure of die casting die steels is not an isolated problem[2]. Ingot moulds, forging dies, hot extrusion dies and railway wheel moulds are affected by thermal fatigue and it is likely that, in the future, industries such as these will cooperate with the die casting industry to investigate how die and mould performance can be improved. The mathematics of die failure by thermal fatigue is complex because many variables have to be taken into account and many assumptions made. No formula has yet enabled a die caster to ascertain precisely how long his dies will last, but the work that has been done indicates that, for one reason or another, die casting dies fail after far fewer shots than the theoretically calculated life.

The factors which affect die life under conditions of thermal fatigue are summarized in the list below.

1. The composition of the steel must make it suitable for being brought to a sufficient hardness and strength to endure the conditions in the die.
2. The steel's physical properties determine the stresses set up under conditions of rapid heating and cooling. Thus a low coefficient of expansion indicates a minimum dimensional change. High thermal diffusivity ensures the quick dissipation of heat.
3. Tensile, compressive and fatigue strengths at elevated temperatures affect the endurance of the steel under stress. The modulus of elasticity is also relevant; this is a measure of the ratio of applied stress and the amount of dimensional change caused by that stress.
4. The steel quality determines whether the die components will have a long or a short life; any inclusion or crack is a stress raiser from which early fatigue failure will be initiated. A well made and carefully finished die is likely to have a longer life than one which has file marks, notches and inclusions. A die which has been damaged in manufacture is liable to deterioration in service, for example the die component shown in *Figures 26.3* and *26.4* which had been welded without proper care.
5. Heat treatment can be the cause of long die life or early failure. Some of the examples discussed at the beginning of the chaper illustrate abbreviated die life brought about by inefficient heat treatment. This essential process must be done under the best possible conditions. If the die casting company does not possess suitable equipment and well trained metallurgical supervision, it is preferable to contract the work to a specialist, and to cooperate with that company. Some die casting managements, surprised by what they considered to be a high quotation for heat treatment, have cut the cost by insisting on short circuiting the processes, but have regretted it later, when a die failed prematurely.
6. The thermal conductivity of the diecast metal has an effect on the transfer of heat. This was discussed in a paper by S.J. Noesen and H.A. Williams[3], who pointed out that the rate at which heat is conducted across the interface of die steel and cast metal is determined by the thermal conductivities of both materials. A die steel with greater conductivity than that of the metal being cast will cool more rapidly than a die into which a metal with higher conductivity is being cast.

The stress set up in a die insert is caused because the temperature of the steel is momentarily increased from about 300°C to a temperature approaching 600°C when the molten aluminium alloy is injected. Tests have indicated that for a fraction of a second the surface of the die is increased to within $10°C$ of the alloy casting temperature. Then, as the metal is rapidly chilled and heat is transmitted through the die to the cooling fluid, the temperature reverts to about $300°C$. The bolster which holds the die insert suffers very little temperature change during the die casting cycle. The coefficient of thermal expansion of H.13 within the operating range of temperatures is approximately $14.6 \times 10^{-6}/°C$. If a die insert 100mm across were allowed to expand and contract freely on a temperature increase of $300°C$ the dimensional change would be about 0.44mm, an amount greater than the extension caused by exceeding the yield point of the steel. Since the expansion of a die surface by a similar temperature increase is restrained by the mass of the die, plastic deformation occurs. On subsequent cooling, the skin is brought under tension and continued cycling will eventually lead to craze cracking.

If the stresses were applied only once, failure would not occur but the continuous repetition of stresses ultimately cause the die steel to develop craze cracking. These will grow at an accelerating rate after a number of die casting cycles. As with any stress raiser, the first crack is the 'beginning of the end'.

It is often possible to plan the die design so that worn parts can be replaced. Die gate areas and projecting cores in line with the stream of injected metal are the first to show craze cracking because they are subjected to the highest thermal gradients. In addition, such areas suffer from the effect of erosion. Where possible, metal fed into a cavity is directed around any protruding core and along the casting, rather than across the wall. A separate replaceable insert pad can be fitted around the gate areas. Often, on symmetrical castings, the die gate can be recessed into a separate insert ring. When cracking begins to advance past the ring on to the insert cavity the ring can be turned, enabling metal to be directed into the cavity from a different direction.

One of the most important contributions to the literature on die thermal fatigue was made in 1970 by Benedyk, Moracz and Wallace[4]. They employed a test which simulated the thermal cycle in die casting aluminium alloys. A number of steels and heat resisting materials were compared with the 5% chromium H.13 steel which, by then, was becoming widely used as a die insert material. They recommended that in heat treatment the austenitizing temperature should be $1070°C$ − an increase from the figure of $950°C$ which had been customary before. They also proposed that the maximum resistance to thermal fatigue in quenched and tempered H.13 steel was obtained when the hardness was 48 HRC. Heat treated steel of lower hardness was insufficiently strong, while greater hardness led to brittleness. It must be added that now an austenitizing temperature of $1050°C$ is considered better than $1070°C$. The hardness of the die inserts required for aluminium is decided according to the size of insert and the anticipated working conditions. In the past, following the recommendations mentioned above, many die casters hardened all their die inserts to 48 HRC, only to find that early fatigue failures occurred. Today, in addition to making large inserts in several pieces to confine fatigue failures to limited parts of the die, hardness levels are chosen according to the size of the die insert, 42−44 HRC for large inserts, increasing to 46−48 HRC for small ones. In the field of zinc alloy die casting, where the dangers of thermal fatigue are much less than with aluminium, die hardness as high as 50−52 HRC is sometimes used. In one mathematical study[5], the die life of a steel under conditions of thermal fatigue was shown to be dependent on the tensile strength at operating temperature, divided by the product of elastic modulus and temperature difference between metal and die, all to the eighth power. Such a formula indicates that a die operating at a temperature range of 300°C would have 10 times the life of a die operating within a range of 400°C. This emphasizes the fact that careful temperature control and the smallest practicable difference between die and casting temperature is the key to maximum die life.

The paper by Noesen and Williams[3] referred to on page 270, gives a number of formulae relating die lives with operating conditions, mainly with molybdenum inserts for die casting in copper and stainless steel but their findings are relevant to the behaviour of dies for aluminium. Most of their formulae are complex and involve making several assumptions but they give one formula which can be simplied to relate the number of 'cycles to failure', N, with 'cyclic strain', S,

$$N = \left(\frac{C}{S}\right)^2$$

were C is a constant for the die material, related to ductility in tension. Using very basic calculations, taking an arbitrary constant of 10, the die lives for cyclic strains of 0.05, 0.1 and 0.2 would be as follows;

die life

(a) $\left(\dfrac{10}{0.05}\right)^2$ = 40 000

(b) $\left(\dfrac{10}{0.1}\right)^2$ = 10 000

(c) $\left(\dfrac{10}{0.2}\right)^2$ = 2500

This emphasizes that a comparatively small increase in cyclic strain leads to a drastic reduction in die life.

Those who are responsible for testing dies do not always realize that the failure of a die steel is connected with the temperature difference during the cycle, and that a die which has been allowed to cool too much before the next injection of metal has received just as much stored up damage as an overheated die. Several papers presented during recent years have emphasized the need for careful heat treatment[6], good die casting practice[7] and stress relieving[8].

Preheating dies before production commences often receives inadequate control. Temperatures of critical die parts such as large inserts and protruding cores should be monitored to ensure that they have reached a sufficiently high temperature to reduce thermal shock on metal injection. When cooling is required, the sudden flow of the water will also lead to thermal shock; it is good practice to maintain a low flow of cooling fluid even during die preheating. When the production cycle is halted for more than a few minutes, a solenoid valve situated on the machine can be sequenced to control coolant flow.

The practice of stress relieving during the course of production is still not universal, but die casting producers are discovering that improved die life can be obtained. The Swiss company, Buhler Brothers, who manufacture machines and die castings claim that they obtain die lives over the 70 000 mark. Some comments about stress relieving are contained in one of their technical booklets[9], and in their training courses they emphasize the benefits to be obtained. A typical routine involves stress relieving after 500, 5000 and 25 000 shots. Some companies also insist that after spark erosion of a die component it should be rehardened.

Although all die casters and nearly all users are aware of the production benefits to be obtained by 'designing for die casting', far fewer are informed of the equally great benefits obtained by 'designing for long die life' and 'designing for effective heat treatments'. All the evidence from theoretical studies indicates that the die insert for an aluminium die casting should be capable of providing longer life than the present average. The required composition and quality of the steel is well understood; heat treatment is generally carried out under precisely controlled conditions; methods of testing the quality of the die steel and of the finished die are well established. The treatment of the die in the foundry can make or mar the performance of the die but this is controlled with far less precision than the other factors. Toolmakers are often appalled at the way in which their expensively produced dies are treated in the foundry. Water hoses being turned on the dies to obtain more rapid production are not so much seen now as they were 15 years ago, but the attitude of 'rate of production at all costs' is still prevalent. One bad habit that is not given much publicity, because it is not much noticed, is the overheating or under cooling of dies during meal breaks. It has sometimes been suggested that if die casting supervisors could be given training in the toolroom and if toolroom supervisors were required to work in the foundry for a period there would be greater appreciation of die life problems.

References

1. BNF Metals Technology Centre
    An analysis of high pressure diecasting die failure. *BNF report.* RRA 1958, Wantage, Oxfordshire (July 1979)

2. COFFIN, L.F.
    Thermal stress and thermal fatigue. *Society of Experimental Stress Analysis,* **15,** (2), 117–130 (June 1957)

3. NOESEN, S.J.; WILLIAMS, H.A.
    The thermal fatigue of diecasting dies. *Modern Casting,* **51,** (6), 119–132 (June 1967)

4. BENEDYK, J.C.; MORACZ, D.J.; WALLACE, J.F.
    Thermal fatigue behaviour of die materials for aluminium die casting. *Society of Die Casting Engineers* Congress Paper 111 (1970)

5. MANSON, S.S.
    Thermal fatigue of die casting dies – a review and proposed future program. *Die Casting Research Foundation.* 01-72-05D (1972)

6. JADHAU, U.; SESHAN, S.; MURTHY, K.S.S.
    Heat treatment of die casting dies. *Tool Alloy Steels,* **12,** (12), 375 (December 1972)

7. YOUNG, W.
    Why die casting dies fail *Society of Die Casting Engineers* Congress paper GT-79-092 (March 1979)

8. YOUNG, W.
    Die casting die failure and its prevention. *Precision Metal Molding,* **26,** 38 (March 1979)

9. KEIL, E.; KOCH, P.
    The development of a die. *Technical brochure No. 6* Buhler Brothers Ltd, Uzwil, Switzerland

Chapter 27

Heat treatment of die steels

Although alloy steels contain elements such as chromium, molybdenum and vanadium, two constituents are essential for heat treatment: iron, termed ferrite in metallography, and carbon, which combines with iron to form cementite, the hard intermetallic compound Fe_3C. These two constituents form a eutectoid structure known as pearlite when the steel is cooled slowly enough to reach equilibrium, but by rapid cooling the steel is hardened. When such a quenched steel is tempered, structures with mechanical properties intermediate between those of the slowly cooled and the quenched conditions are formed.

The iron–carbon equilibrium diagram for steels up to 1.1% carbon in *Figure 27.1* shows the temperatures of the boundaries within which the phases are stable[1]. The upper and lower transformation temperatures are given the suffix c when the equilibrium diagram denotes heating and suffix r for cooling. At ambient temperature, an 0.3% carbon steel that has been cooled slowly consists of ferrite and pearlite. After heating to the A_{c1} lower transformation temperature of $723°C$, carbon enters into solid solution to form the austenite phase, and at the A_{c3} upper transformation temperature of about $825°C$, the steel is fully austenitic. Slow cooling from this condition causes the structure to revert back to ferrite and pearlite, their ratio under equilibrium conditions being dependent on carbon content. Structures A and B in *Figure 27.1* illustrate a decrease in the ferrite/pearlite ratio with increase of carbon. The steel is entirely pearlite (view C) at the eutectoid composition of about 0.8% carbon. Above that amount the structure consists of pearlite with cementite at the grain boundary regions (D). On rapid cooling from the austenitic region, the carbon diffusion mechanism, which is temperature and time dependent, is not allowed to occur and the martensite which is formed distorts the atomic lattice, causing hardening and strengthening, depending on the severity of quench.

In recent years there has been a greater understanding of the complex structural changes taking place during heat treatment, with the help of phase transformation diagrams. Use of these diagrams can lead to better control of the heat treatment cycle which in turn will ensure that optimum properties and maximum die life are achieved.

Isothermal transformation diagrams, sometimes referred to as time-temperature transformation (TTT) or S-curves[2], describe the effects of cooling from the autenitizing temperature for a given steel specification and allow for any alloying elements that are present. Curves for a 5% chromium die steel are shown in *Figure 27.2*. By comparing a fast cooling rate, for example a water quench along path (A) with a slower rate typical of

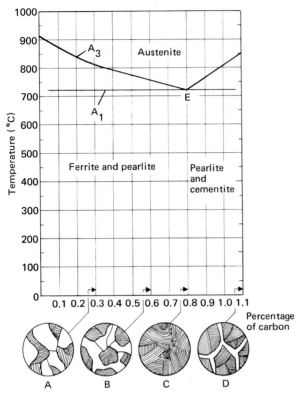

Figure 27.1 Iron-carbon equilibrium diagram, showing transformation temperatures in steel

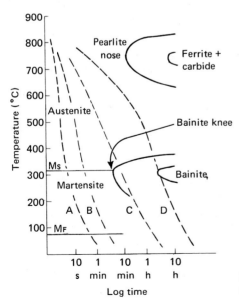

Figure 27.2 Isothermal transformation diagram for hot die steel BH 13

an oil quench along path **(B)**, the steel would transform from an austenitic structure to a fully martensitic structure, commencing at about 300°C since no other transformation lines are crossed. A slower cooling rate along path **(C)**, typical of air cooling, would produce an intermediate structure, bainite, between 330°C and 210°C after which the remaining austenite would transform to martensite. Alloying additions of chromium and molybdenum displace the transformation boundaries to the right of the diagram to confer additional hardenability. Low alloy steels do not possess such hardenability and only fast cooling rates along path **(A)** typical of a water quench would produce a fully hardened structure. Slower rates of cooling would be accompanied by some transformation to softer structures.

The rate of cooling on the surface of a die will be much higher than the centre section. It is, therefore, necessary to take into account the variations in hardness that will be expected throughout the die section. Reference can be made to continuous cooling transformation diagrams[3], known as CCT curves, shown in *Figure 27.3* which equate the

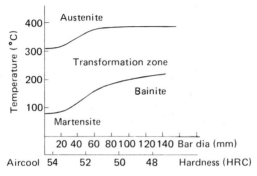

Figure 27.3 Continuous cooling transformation diagram for hot die steel BH 13

levels of hardness expected for a given die section thickness. The production of each CCT curve requires the measuring of a large number of cooling rates, using a computer controlled furnace, and the microexamination of all the structures obtained. Often steel manufacturers make themselves responsible for the experimental work leading to the publication of each new curve. As more and more CCT curves are produced, their use increases.

Stress relieving

The H.13 steel supplied for die manufacture is normally in the soft annealed condition. Machining of a die, particularly when large amounts of metal are removed, induces stresses which may cause distortion or warpage during heat treatment. Sharp corners and small radii on a die cavity are undesirable since they act as stress raisers during heat treatment and in service life afterwards. After rough machining the die to 3 mm over finished size, it is stress relieved by heating between 620°C and 700°C and held at this temperature for 12 minutes/cm of greatest die cross section or for 2 h, whichever is the longer[4]. Air cooling from this temperature is normally acceptable, but where there are large changes in die cross section, slow cooling in the furnace at 425°C or lower, to minimize thermally induced stresses, is recommended. Any distortion resulting from the release of stress can be corrected before final machining.

Additional stress relieving of dies at predetermined intervals during their production can extend die life by retarding the onset of fatigue cracking. A system of regular stress relieving is operated by some die casting companies who claim substantial increases in die life. The die should be stress relieved at a temperature of 25°C lower than the previous tempering temperature. If the value of the latter is not known, 540°C is suitable for H.13 steel, holding at this temperature for a time depending on the section of the die.

All die areas where spark erosion techniques have been used will have a thin hard surface layer containing micro-cracks which can lead to early thermal fatigue failure. Although an established technique is to remove this hard surface by machining or hand working, it is not always practicable on deep and complex die shapes and stress relieving is recommended.

Preheating

In preparing the die steel for hardening it is important to preheat it in stages, in a controlled atmosphere furnace to a temperature of approximately 850°C, just below the austenitic transformation region. Control in preheating ensures that when the austenitizing temperature is eventually reached, the surface and centre sections of the die undergo the transformation volume change at the same time, to reduce stress and distortion.

The dilatometric curve in *Figure 27.4* shows the volume changes which occur during the full heat treatment cycle of a die steel. Expansion of steel is uniform during heating

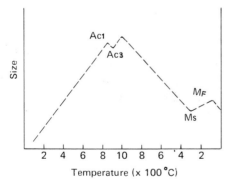

Figure 27.4 Schematic diagram showing size changes during hardening of a die steel

to the start of austenite transformation at A_{c1} (approximately 850°C) but contraction then occurs due to the change from a body-centred lattice of ferrite to a face-centred lattice of austenite until this transformation is complete at A_{c3} (approximately 890°C). A further expansion occurs when heating up to the hardening temperature. Cooling from this temperature shows a uniform contraction until the start of martensite formation shown as M_s on *Figure 27.4* (M_f signifies the finish at the transformation). Martensite formation is accompanied by a tendency to expand, the degree of expansion depending on the alloy content of the steel and in some cases it varies according to direction.

Hardening

Following a controlled preheating cycle, the die is then heated to 980–1040°C, into the austenitic region. This ensures that, in addition to carbon, a high percentage of the alloying elements will be taken into solid solution. The temperature used is a compromise dictated by the casting temperatures of the alloys to be cast, and some are given in *Table 27.1*.

TABLE 27.1

	Hardening temperature (°C)
Zinc alloy	990–1010
Aluminium alloy	1000–1030
Brass	1020–1040

The hardening temperature employed will affect the toughness and strength after hardening. High temperatures encourage dissolution of alloying carbides, leading to maximum hardness and hot strength after quenching, but larger austenite grains will grow, reducing the toughness of the steel. Low temperatures produce a tougher steel but, without maximum carbide dissolution, hardness and hot strength will be impaired.

Soaking time within the furnace to achieve a uniform temperature throughout the die block will depend on die configuration, furnace design and rate of heating. Following heating and soaking at the required temperature the steel is quenched in order to retain the maximum amount of alloying elements in solid solution. Although H.13 is classed as an air hardening steel, larger die sections require more severe quenching in order to achieve optimum properties. In practice some of the soft austenite phase may become sufficiently stable and remain untransformed. This retained austenite is found in steel where segregation has occurred and bands rich in carbon have formed.

Tempering

The presence of martensite and bainite in a quenched steel, whilst increasing tensile strength and hardness, cause the material to be brittle. By a controlled tempering treatment, some of the residual stresses in the quenched steel are relieved by allowing carbon in the strained lattice structure to precipitate as a finely dispersed phase, usually described as 'tempered martensite'. The toughness of the steel is improved with little effect on hardness and tensile properties. Tempering also encourages transformation of any retained austenite to secondary martensite which requires to be stress relieved by retempering.

To ensure that residual stresses after hardening are not allowed to cause premature cracking or distortion, it is good practice to begin tempering treatments before the steel has cooled to room temperature but not until the steel is below 50°C. A first tempering temperature within the range of 560–600°C is normally specified. The die steel must not be loaded into a furnace hotter than 250°C initially, or alternatively, the steel is preheated to 250°C before loading into a furnace at the tempering temperature. A soaking treatment is followed by cooling to room temperature. Formation of secondary (transformed) martensite will require to be further stress relieved by retempering the work between 570°C and 640°C, the actual temperature being that which will give the required properties as indicated in *Figure 27.5*. If the required hardness is not achieved, a third temper can be given.

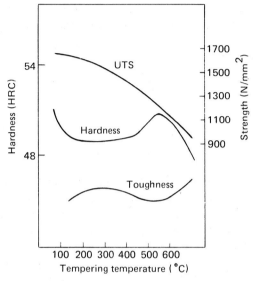

Figure 27.5 Mechanical properties of H.13 hot die steel after tempering

Heat treatment plant

Although heat treatment may account for only a typical 5% of total die cost, it has a major effect on the performance of the die. If the treatment is not carried out properly, the die caster could incur substantial rectification costs or even complete die replacement after a short period of service. Developments in heat treatment technology in recent years have encouraged many die casters to subcontract work to heat treatment specialists. A recent paper by T. Atterbury[5], stresses the advantages offered by outside contractors, and reviews the various types of equipment which are suitable for the heat treatment of alloy steels for tools and dies, together with changes made in the applications and technology of furnace design.

Muffle furnaces

These are the cheapest and simplest furnaces available for die steel treatment and are suited for the die caster who wishes to install a plant for the treatment of cores. Furnaces can be gas heated, either directly or through radiant tubes made of refractory material, or electrically heated. To prevent scaling and decarburizing of the steel surface, protection is given by enclosing items to be treated in a containiner with cast iron swarf, nonactivated charcoal or spent coke. Alternatively, a protective paint or paste, based on Kaolin (china clay) with a wax binder, which is easily brushed off after use, is employed.

A development from the basic muffle furnace uses a controlled atmosphere to protect the die surface and incorporates a sealed quench unit where work can be transferred automatically. *Figure 27.6* represents a semicontinuous sealed quench furnace

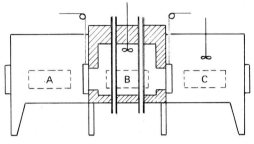

Figure 27.6 A semi-continuous sealed quench furnace with radiant tube heating

where the work is loaded in section A and flushed with protective gas before transfer to the central heating chamber B and quench chamber C, both protected by gas. Protective gases (endogas), are generated by cracking hydrocarbon fuel gas with air over a heated catalyst. These gas atmospheres are predominantly nitrogen-rich, but contain carbon dioxide, carbon monoxide, hydrogen and methane, to provide a controllable carbon potential that will be neutral to the carbon content of the steel being heat treated. At present sealed quench furnaces have been used to only a limited extent for die heat treatment, but it is expected that they will be employed more widely during the coming years.

Salt baths

Die steel treatment in salt baths represents a technique that was well established until ecology problems, and the difficulty and cost of waste disposal, caused its use to decline in favour of more acceptable methods. For temperatures over $950°C$, required for H.13 die steel, refractory pots heated by electrodes are used. The types of salt employed vary depending on the treatment temperatures required. A preheat bath contains a mixture of barium, potassium and sodium chlorides while, for austenitizing, up to 95% barium chloride with 5% of silica or common salt is used. Quenching and tempering baths use chlorides of calcium, barium and sodium.

The salt bath is resistance heated using the salt as the resistor; electrodes set into the furnace walls transfer the current. The magnetic field induced around the electrodes ensures that the molten salt is stirred electrodynamically, leading to uniform bath temperature. Problems associated with these furnaces include the retention of salt on the die steel surface (i.e. as salt is hygroscopic, salt fumes condense and corrode heat treatment plant and other equipment). A further hazard is water pick up in the baths and condensation which can cause an explosion from spontaneous vaporization of the water. An oxidizing potential in the bath which can decarburize the treated steel arises from absorption of oxides from the work and atmospheric oxidation at the salt/air interface which produces barium and/or sodium oxides. Oxide scavengers, termed regenerators or rectifiers, based on silicon (less frequently boron) are required to overcome this problem. These materials encourage precipitation of heavy oxides as a sludge which can be removed by a ladle. Alternatively, silicon tetrachloride or methylchloride is bubbled through the molten salt in a stream of nitrogen gas.

Fluidized beds

The technique of fluidization[6] dates to the end of the 19th century when such furnaces were used for conveying, drying and gasification of coal. Heat treatment in fluidized beds began to be developed in the early 1950s but at first their application was restricted to processes requiring temperatures of not more than 700°C. The advent of fuel fired fluidized beds using gas-air mixtures for both heating and fluidization led to the economic development of the process for heat treatment of steel wire and small components and later to large components and parts of die casting dies.

The 'bed' is an insulated container, filled with inert particles such as alumina or sand. A premix gas-air mixture is blown upwards into the furnace at high velocity through a porous bottom tile; this fluidizes the particles, while the combustion of gases provides the heat. Components to be treated are immersed in the bath of particles as if they were in a liquid. The heat transfer rate in the bed is considerably higher than in an open fired furnace. So far the use of fluidized beds in die casting has been centred around the replacement of salt bath equipment, on account of greater efficiency and environmental advantages. With efficient combustion conditions, a decarbonizing atmosphere is produced. The use of proprietary protective pastes or enclosing the parts in stainless steel foil have been found satisfactory. Modern furnaces are being manufactured with external heating; an inert gas is used for fluidizing the bed.

Vacuum heat treatment

Vacuum heat treatment of die casting die steels owes its beginning to developments of furnaces designed for vacuum brazing and annealing. A next stage came with the advent of high duty nickel-base alloys requiring solution heat treatment under closely controlled conditions. Then the method was extended to die steels[7]. Such furnaces had advantages over other established processes but the early equipment was not always satisfactory, partly because technicians did not understand the new process and partly because the furnaces were not always trouble free. During the 1970s great strides were made, both in the improvement of vacuum furnaces and in the understanding of them by users. Die casting managements realized that such a process, which will result in a better quality die and a longer die life, is worth extended use. As we discussed in the previous chapter, die life depends on many factors, not least the treatment given to the die in the foundry but there is sufficient evidence to justify the use of vacuum heat treatment which can provide a die to go on the die casting machine in the best possible condition[8].

The essential difference between vacuum and other heat treatment processes is the evacuation of the heating chamber during the heating up and holding of the steel being treated[9]. Subsequently the work load is quenched in a neutral gas or in an immersion oil quench bath. The removal by evacuation of any active gases provides neutral conditions in the furnace and prevents the occurrence of oxidation, decarburization, carburization and nitriding. No effluent or dross are produced and no heat is evolved, while the gentle thermal changes achieved ensure that distortion and quench-cracking are minimized. A high surface finish is retained on the treated parts, so that very little final grinding, or none at all, is required. Capital costs are fairly high and the equipment is generally operated by specialist contractors who can provide the necessary technical supervision and facilities. The vacuum used in these furnaces is of the order of $1-500\mu m$. Both vertical and horizontal furnaces are available, although to facilitate charging of die casting die blocks the horizontal type are preferred. A double-walled furnace provides the means to water cool, this feature also being of use to cool the gas quenchant if a separate heat exchanger is not employed.

The vacuum furnace shown in *Figure 27.8* incorporates an oil quench which is an integral part of the vacuum system allowing fast quench of parts and maximum

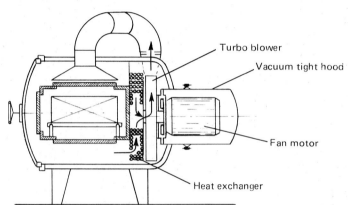

Figure 27.7   Turbo-quench, positive pressure vacuum furnace.
(Courtesy Ipsen Industries International)

hardness in the die steels. The heating position is hermetically sealed from the quench zones. Handling is confined to loading and unloading each batch. Work enters the loading area which is then evacuated to the same level as the rest of the furnace before the work is transported to the heating chamber. The central furnace is for hardening and the one on the right for tempering. Following the heating cycle the work is transferred to the cooling chamber, where quenching can be achieved either by lowering the steel into an oil quench bath or by flooding the chamber with a recirculated inert gas. When oil is used as a quenchant it should be free from components having a low vapour pressure. The oil quench equipment is suitable for contractors who heat treat a range of alloys, non-ferrous and ferrous.

Companies specializing in treatment of die steels find gas-quench furnaces *(Figure 27.7)* more suitable. Heating is normally achieved with graphite resistance elements. Gas quench media used, in reducing order of efficiency, are hydrogen, helium and nitrogen; plentiful supplies of high quality nitrogen make this gas the obvious choice against the high cost of helium and the inherent dangers in the use of hydrogen.

Figure 27.8 Oil quench two-chamber vacuum furnace. (Courtesy Ipsen Industries International and Aldridge Tool Heat Treatment Ltd.)

Design for heat treatment[10]

Due to the change in dimensions of a steel during heat treatment, the ideal die would have a uniform cross section and absence of thin rib sections, holes or sharp angles. In practice this is rarely possible, because dies are required to produce castings of complex shapes, but dies should be designed to permit heat treatment to be as efficient as possible. The die casting industry advocates that intelligent component design will assist casting manufacture and finishing operations. What has not been emphasized clearly enough is that early cooperation between the die designer and heat treater will help to avoid some of the problems which can arise during heat treatment and subsequent die casting production.

Thin die sections on inserts and cores are typically vulnerable to distortion and cracking and could be made as fastened-in inserts and cores which are easily replaced. Alternatively, these areas could be left oversize for heat treatment and machined to finished size before casting production. One simple precaution is sometimes overlooked by die casting companies which have embarked on the design and manufacture of a die much larger than they have ever tackled before. They should always make certain that plant is available to treat the size of dies being designed.

References

1.   ALEXANDER, W.O.; STREET, A.C.
      Metals in the service of man, Eighth edition, Chapter 11 Penguin Books Ltd. (1982)

2.   PETER, W.; HASSDENTEUFEL, H.
      Information value of the end quench test and test-temperature transformation diagram. *Stahl und Eisen,* **87,** 55 (April 1967) (Available as translation. BISI. 5644)

3.   ATKINS, M.; ANDREWS, K.W.
      Continuous cooling transformation diagrams. *British Steel Corporation,* Report SP/PTM/6036/71/C

4.   *Code of practice for heat treatment of 5% chromium die steels (H. 13 type)* Zinc Alloy Die Casters Association, London

5.   ATTERBURY, T.
      Heat treatment plant for tools and dies. *Metallurgia,* **47,** (9), 438 (September 1980)

6.   FENNELL, A.G.; VIRR, M.J.
      Fluidized bed heat treatment. *Heat Treating Magazine,* **10,** (12), 16 (December 1978)

7.   REYNOLOSON, R.W.
      Vacuum heat treatment of diecasting dies *Diecasting and Metal Moulding,* **2,** (9), 15 (January/February 1970)

8.   ATTERBURY, T.
      Tool and die heat treatment: a user's assessment of available equipment. *Heat Treatment of Metal,* **2,** 87 (1975)

9.   LUITEN, C.H.; KRIA, E.
      Vacuum heat treating. *Ipsen Industries International Technical Report 22/E*

10.   (Editorial)
      Heat treatment: an appraisal of current trends. *Metals and Materials,* **28** (January 1975)

Chapter 28

Surface treatments for steels

During the past 20 years, several processes have been introduced to obtain enhanced surface hardness of steels. Some of them have developed from case carburizing and nitriding, to obtain faster processing times with better environmental control and improved properties. Various salt bath processes[1,2] have been used and now a wide range of new methods is available[3].

In the die casting industry surface treatments are applied to steels to improve the properties of nozzles, ejector pins, cores and shot sleeves, to provide maximum resistance to erosion, pitting and soldering. Treatment of die cavities has received only limited acclaim, because the complex thermal patterns produced on large die components lead to stresses which are sufficiently high to break through the thin surface treated layers, leading to premature failure. Experience in drop forging has also indicated that surface treatments of their dies have not been particularly successful.

In the United States, research sponsored by the Die Casting Research Foundation, the technical arm of the American Die Casting Institute, is being done at Case Western Reserve University. This work studies the effects of different surface treatments on the performance of H.13 die steel.

A program is also underway that is funded by the National Aeronautics and Space Administration (NASA). The ADCI-DCRF, in conjunction with Case Western Reserve University is studying the effect of ion deposition of protective materials to die surfaces. Upon the conclusion of this testing to determine the most promising material, the process will be studied by using a two-cavity production die with one cavity treated and the other untreated for control purposes.

Thermochemical treatments are applied to die casting die components; the surface chemistry of the steel is modified by the introduction of nitrogen, carbon and sometimes other elements; the processes are of the main types listed below.

1. Nitriding.
2. Nitrocarburizing such as Tufftride, Sulfinuz and Sursulf.
3. Metallizing such as boronizing and the Toyota diffusion process.
4. Carburizing and carbonitriding.

Further classification can be given according to the media used which include gaseous, salt bath, vacuum and pack processes. Electro or electroless metal plating and spray/dip coating are used only occasionally in die casting.

Nitriding

The nitriding process uses a diffusion-reaction principle to increase the surface hardness of steels that have already been hardened and tempered to the required core strength. Aluminium, chromium and vanadium have strong tendencies to form nitrides, so diffusion of nitrogen into a steel containing one or more of these elements produces nitrides. The fine dispersion of these particles confers surface hardness. Nitriding is commonly used for the hardening of ejector pins. Those made of chromium—vanadium or H.13 are hardened, tempered and ground before being nitrided or treated with newer processes which will be described later. Nitriding has found some application in the hardening of cores although the Tufftride and Sursulf processes offering higher fatigue and wear resisting properties are often preferred.

Nitriding of cold chamber machine shot sleeves to a depth of about 0.25 mm increases the surface hardness to about 70 HRC. Since it is cheaper to replace a worn plunger tip than a complete shot sleeve, it is preferable to maintain shot sleeves at higher hardness levels than the plunger tips to reduce wear and prevent eventual seizure. However, as discussed on page 197 beryllium—copper and other copper alloy plunger tips are being used to an increasing extent, avoiding the need for sleeves to be nitrided.

Nitrogen is introduced into the surface of the steel by reaction with a gaseous or liquid phase, or, more recently, by nitrogen-containing plasma.

Gas nitriding

Components for gas nitriding are placed in a sealed retort housed in a furnace shell, and heated to about 525°C. Ammonia passed through the retort chemically degrades to yield nitrogen which is absorbed into the surface of the steel. The gas from the retort is monitored for ammonia and the flow of inlet gas is adjusted to give 80—85% ammonia at exit. If the amount of ammonia is too great, the flow rate is reduced until under steady state conditions, the residual ammonia level is reduced to about 50%. Several days of treatment are required to produce a 0.3 mm depth of hardened case.

Liquid nitriding

The liquid nitriding process is carried out in a salt bath containing a mixture of salts including sodium cyanide. At operating temperatures of about 570°C the sodium cyanide (NaCN) oxidizes to produce the unstable compound sodium cyanate (NaCNO), which donates nitrogen to the steel. The technique is several times faster than the gaseous method but is expected to be phased out in favour of the newer processes which do not have the health and safety hazards associated with the disposal of cyanide waste.

Plasma nitriding

Plasma nitriding, also known as ion nitriding, is a comparatively recent development in Britain though it has been used elsewhere for several years. There are no effluent problems of cyanides and the technique is capable of precise control. The components to be nitrided are contained in a vacuum vessel with a negative potential of about 1000 volts between the component (cathode) and the wall of the furnace (anode). The process is carried out using either nitrogen or mixtures of nitrogen with hydrogen which are introduced to a pressure of 0.1–10 m/bar (10–1000 Pa). Under this potential difference, the gas is ionized, producing the luminous phenomenon known as glow discharge, and accelerated towards the workpiece with high kinetic energy. Release of this energy heats the workpiece which is normally controlled to a temperature between 400°C and 600°C though it has been demonstrated that plasma nitriding can take place at temperatures as low as 350°C. The absorption of nitrogen into the steel surface is a function of the electrical voltage and current used, for equal treatment temperatures the time required for the process is only about 30% of that for gas nitriding. During treatment, steels with tenacious oxide films, such as those containing chromium, are cleaned and surface depassivated ready for the nitriding reactions. In view of the use of chromium in die steels, this effect provides an important advantage of plasma nitriding over other processes. The compound layer produced is less susceptible to spalling, thus eliminating the need for further surface finishing. A further benefit of the process is the low temperatures which can be employed leading to less distortion and deterioration of the surface finish than occur in other processes.

A die in H.13 steel for the production of aluminium rotor castings, plasma nitrided for 36 hours at 510°C in a 25% nitrogen, 75% hydrogen gas mixture at a pressure of 10 m/bar (1000 Pa) resulted in a 0.2 mm nitrided case depth with surface hardness values in excess of 63 HRC. Put into service without any further finishing operation, there was a reduced tendency for aluminium to solder to the die and savings were made due to less down time and repolishing costs[4].

Nitrocarburizing

The principle of nitrocarburizing is the introduction of both nitrogen and carbon to form a compound layer of nitrides and carbides in the steel surface. Some of the processing techniques also use small amounts of sulphur compounds which confer increased scuffing, wear and fatigue resistance to the steel. Tufftriding, Sulfinuz and Sursulf processes are liquid salt treatments while more recently gaseous and plasma processes have been developed. Treatment temperature for each method is commonly 570°C and, as with nitriding, the steels are hardened and tempered before treatment to ensure optimum properties.

Tufftriding

This is a proprietory liquid nitrocarburizing process; a mixture of potassium cyanide and cyanate is held in a titanium-lined bath at about 570°C. Carbon and nitrogen are derived from the decomposition of the cyanate which in turn is obtained by the oxidation of the cyanide salt, accelerated by air bubbling. Nitrogen diffuses into the treated components while iron carbide particles at or near the surface, act as nuclei to precipitate some of the diffused nitrogen as carbon-bearing iron nitride. Treatment time to produce a compound layer of 0.01–0.02 mm is 1–2 hours, and the process is comparatively economical.

None of the brittle iron nitride is formed as in gas nitriding whilst improvements in fatigue and endurance properties are due to the diffusion zone where nitrogen in solid solution forms a barrier which helps to prevent incipient cracks from becoming fatigue failures. To maintain the nitrocarburizing activity of the salt bath mixture, spent salt and reaction sludges need to be removed and fresh cyanide salt added.

Following attempts to reduce the downtime caused by these operations and create better environmental and safety conditions arising from the use of cyanide salts, a new Tufftride process[5] was recently developed using sodium and potassium cyanate and carbonate salts. A small amount of cyanide free regenerator reacts with the carbonate forming fresh cyanate to maintain a level of about 36% cyanate in the bath. A low cyanide concentration develops but normally remains below 3%. Using the new process, cost savings are reported and there are practically no reaction sludges for disposal. Treatment times are short and the same depth of compound zone can be achieved in less than half the normal treatment time. Even Tufftrided parts with a compound zone depth of only 0.003 mm often demonstrate greater wear resistance than similar components which have been nitrided to higher hardness and much greater case depths. This process has proved successful in applications to die areas where lubrication is difficult or impossible.

Sulfinuz

The Sulfinuz treatment is also a liquid nitrocarburizing process carried out in a bath containing salts of sodium cyanide, cyanate, carbonate and sulphide at 570°C. It can be applied to all ferrous materials and the mechanism and depth of case produced is similar to Tufftriding except that simultaneous diffusion of sulphur into the surface occurs to give greater resistance to fatigue and wear.

Sursulf

This liquid process was developed to produce a hard wear resistant surface layer on steels without any of the problems of effluent disposal associated with the Sulfinuz treatment[6]. The bath is based on a eutectic mixture of cyanate and carbonate salts of lithium, sodium and potassium with a small amount of potassium sulphide. The carbonate produced, instead of being allowed to build up as a wasteful product of the reaction, is recycled by conversion back to cyanate by means of cyanide-free regenerators, to provide the source of nitrogen.

Metallizing

Metallizing and similar processes are used to deposit metals or elements having both metallic and non-metallic properties onto the surface of steel. High surface hardness and resistance to wear are the main properties achieved but to date only limited use has been made of the processes. Treatment temperatures are usually in excess of 900°C and precautions need to be taken to minimize distortion.

Boronizing

A diffusion process analagous to carburizing is boronizing[7]. Following the deposition of boron to a steel at about 900°C, iron borides are formed conferring high surface hardness which measured on the Vickers scale is in excess of 1600 HV. Pack and gaseous processes are generally used. Treatment times up to 6 hours at 900°C produce a surface layer of 0.15 mm on carbon and low alloy steel, while thinner layers up to 0.02 mm would be adequate for high alloy tool steels which are hardened and tempered after treatment for optimum core properties.

Toyota diffusion (TD) process

This proprietory diffusion coating process was developed in Japan and is being introduced to other countries. It has been applied to most tool steels and to tungsten—carbide—cobalt materials and produces adherent surface carbide[8] layers 5—15 μm thick which are satisfactory for most applications. Titanium, vanadium, niobium and chromium carbides have been produced with hardness levels of about 3000—4000 HV for titanium and vanadium carbide layers, 2800 HV for the niobium carbide layer and 1700 HV for the chromium carbide layer. The very high hardness is associated with wear resistance greater than that of nitrided steels. The treated steel is tough and offers better resistance to soldering, thermal and mechanical shock, heat checking and oxidation than untreated die steel.

The process is carried out in a salt bath furnace consisting of a heat resisting steel pot with heating elements. The salt bath is of proprietory composition based on borax to which the carbide-forming elements are added, usually as ferro-alloys such as ferro—vanadium, but the type of carbide can be changed easily. A uniform and smooth layer of carbide will form on the die surface by a reaction between the carbide-forming elements dissolved in the borax and the carbon in the base metal to provide a metallurgically bonded structure.

The bath temperature selected conforms to the hardening temperature of the die steel, for example, up to 1050°C for H.13. Immersion times of 4—8 hours are required to produce carbide layers of 5—10 μm on this steel. It is usual to quench dies directly from the bath in water, oil, salt or air depending on the type of steel being treated and the hardness required. Due to the high treatment temperatures, any problems of distortion may be controlled by prehardening or tempering, thus minimizing dimensional changes during the treatment. The process does not give rise to pollution problems and no special treatment facilities are required.

In the Japanese aluminium die casting industry, the coating most widely used has been vanadium carbide; some die inserts and large cores have been treated successfully, but the main application has been for small cores. Over 90 000 core pins have been treated, which, compared to nitrided parts, have given double the service life when located near the die ingate and still more in areas remote from the ingate. Cost savings claimed have been related to prolonged die life and reduced die repair.

Carburizing and carbonitriding

Where certain die components do not require expensive die steels the use of low alloy steels is quite normal and their surfaces can be carburized or carbonitrided for added strength and abrasion resistance. Gaseous or liquid methods are generally used at temperatures of 800—900°C. Diffusion of carbon into the steel in carburizing produces a high carbon surface and, when quenched, the surface becomes hardened. The carbo-nitriding introduces carbon and nitrogen simultaneously into the surface of the steel.

References

1.   TAYLOR, E.
       Improving Wear and Thermal Fatigue Properties of Die Steels. *Precision Metal,* **26,** 38 (August
       1968)

2.   GREGORY J.C.
       Improving the Resistance of Ferrous Metals to Scuffing, Wear, Fretting and Fatigue. *Metal
       Forming,* **35,** 229, 258, 294 (August/September/October 1968)

3.   CHILD, H.C.
       Surface Hardening of Steel. *Engineering Design Guides No. 37* Oxford University Press (1980)

4.   EDENHOFER, B.
       Physical and Metallurgical Aspects of Ion Nitriding. *Heat Treatment of Metals,* **1,** 23, (1974)

5.   ASTLEY, P.
       Tufftride – A New Development Reduces Treatment Costs and Process Toxicity. *Heat Treat-
       ment of Metals,* **2,** 51 (1975)

6.   GREGORY, J.C.
       Sursulf – Improving the Resistance of Ferrous Materials to Scuffing, Wear, Fretting and Fatigue
       by Treatment in a Non-Toxic Salt Bath. *Heat Treatment of Metals,* **2,** 55, (1975)

7.   BIDDULPH, R.H.
       Boronizing. *Heat Treatment of Metals,* **1,** 95 (1974)

8.   ARAI, T.
       Carbide Coating Process by use of Molten Borax Bath in Japan. *Heat Treating,* Vol. 1, p. 15.
       American Society for Metals (1979)

Chapter 29

Developments in finishing processes

Although many die castings, especially those in aluminium alloys, are used in the as-cast condition, there are often circumstances which require finishing processes; these may be grouped according to their purpose as listed below.

1. To protect the castings from corrosion.
2. To act as a base on which colour finishes can be applied.
3. To enhance the decorative appearance of the components, either by a plated or colour finish.

Finishes for zinc alloy die castings

In humid conditions, white corrosion products are formed on the surface, but effective protection can be given by a chromate passivation treatment such as the UK Ministry of Defence specification DEF 130[1]. Following cleaning and pretreatment the castings are immersed in a slightly acid solution containing chromium salts, which forms a dull yellowish film. The process is also recommended as a pretreatment for painting or organic colour finishes.

Electroplating of zinc alloys

Freshly polished zinc alloy die castings have the appearance of chromium plate but soon tarnish in the atmosphere. When a plated finish is desirable but nickel and chromium plating is too expensive, a recently developed process enables parts to be plated directly with zinc[2]. This provides a low cost finish suitable for some hardware and toys which are not required to endure severe corrosive conditions.

Ever since zinc alloy die castings began to be used, efforts have continued to improve and perfect the nickel and chromium plated finish. At an early stage, it was realized that plating solutions suitable for brass were not effective for zinc alloy, and over the years many processes have been developed, in the copper undercoat, the nickel plate and the final treatment with chromium, leading to the wide acceptance of plated zinc alloy die castings for such parts as automobile hardware. The initial coating with copper is important and it must have a thickness of at least 8μm, or 12μm for complex castings.

Nickel is deposited up to about $25\mu m$ thick and chromium up to about $3\mu m$. Standards of thickness are laid down in BS 1224.

In preparation for plating, it is usually necessary to remove flash lines and evidence of the area where the gate has been trimmed away. Automatic polishing processes are replacing manual polishing when production volumes are sufficiently large. The parts to be processed are mounted in jigs and either a single head follows the contour of the casting or several indexing heads each polish selected areas of the surface. For small castings, barrel polishing or vibratory polishing can be used. The parts are mixed with plastic cones, impregnated with fine abrasive particles, in a detergent solution containing corrosion inhibitors. To obtain the best possible finish, components should be plated soon after removal from the polishing machine.

Copper is usually deposited in two stages. An initial deposit is applied from non-levelling electrolytes based on copper cyanide, dissolved in excess sodium or potassium cyanide. Many proprietary treatments follow the preliminary coating with bright levelling acid solutions containing copper sulphate and sulphuric acid. This technique is appropriate for smoothing over surface imperfections after the initial deposit of copper from a cyanide electrolyte. An alternative electrolyte which is non-corrosive to zinc is a weakly alkaline pyrophosphate copper formulation. Although it does not possess such a good levelling ability as the acid solutions, it is only necessary to apply a thin initial cyanide copper film.

The earliest electrodeposited nickel coatings were produced from dull non-levelling nickel electrolytes followed by chromium. Bright nickel levelling solutions containing small amounts of additives, including sulphur, were developed but neither of these gave full protection against corrosion. It was concluded that the poor coating performance was due to local galvanic action being set up at the base of small defects extending through the brittle chromium to the nickel electrodeposit. Special formulations were then developed, leading to the modern duplex nickel coating process which gives optimum combination of properties. A semi-bright sulphur-free nickel coating with high levelling properties is deposited, followed by an additional relatively thin bright nickel coating containing sulphur. Added corrosion protection is due to difference in electrochemical potential between the two layers of nickel. Any advancing corrosion tends to proceed laterally when reaching the semi-bright nickel layer and penetration to the base metal is delayed.

Chromium plating

Many proprietory solutions have been developed during the past 30 years to improve the corrosion resistance of chromium plating. 'Crack-free chrome', introduced in the late 1950s gave excellent results in static tests but castings subjected to vibration corroded rapidly[3]. A 'Micro-crack chrome' having up to about 80 fine cracks to the mm was successful in dispersing the centres of corrosion; a thicker layer was required than for previous processes and this causes some loss of reflectivity but, without affecting corrosion resistance, it is possible to reduce the nickel layer by up to 25%.

Other systems include 'Duplex chrome', where micro-cracked chromium is deposited on to a crack-free chromium layer. A patented duplex nickel process[4] adds a third stressed nickel deposit up to 1.25 μm thick, which encourages the covering layer of chromium to become micro-cracked. In another process, 'Microporous chrome', a second nickel coating contains insoluble non-conducting particles which produce microscopic pores in the subsequent chromium layer to give improved corrosion resistance.

A research programme in the USA investigated the influence of cathodic passivation treatments using sodium dichromate. Increased resistance to corrosion was reported from a reduced electropotential between the nickel and chromium layers[5]. Research work in the UK has led to commercial success in using trivalent chromium baths which are reputed[6] to have five times the plating rate of conventional hexavalent chromium baths, but they produce a darker colour of electroplate.

Chromium plating direct onto other metals has been used for many years, to provide hard coatings at least 0.03 mm thick. Chromium can be plated directly on to zinc die castings to provide a surface with a hardness of about 800 Brinell (75 Rockwell C). This gives hardness, resistance to abrasion, a low coefficient of friction and improved resistance to corrosion. A paper by J.R. Nicolai and R.E. Marce[7] gave information on the application of hard chromium finishes to small zinc alloy die castings enabling them to be used in many applications which had previously been served by other materials.

Painting

Various paint finishes are applied to zinc die castings, usually after treating the parts in a phosphating or chromating solution. Where adhesion is not critical in some cheaper parts, acrylic paints containing an acid etching ingredient are applied after degreasing without any pretreatment. Paints based on epoxy resins or amines which need to be stoved are recommended where corrosion resistance of the parts is essential. A dense and uniform film can be achieved using electropainting techniques where the parts are made cathodic relative to the steel paint tank and coated with specially formulated paints based on water-soluble resins.

The surface of a die casting can be given a textured finish by etching or otherwise treating the die so that the texturing is reproduced on the casting. Such treatments are attractive in their own right or they can serve as a base for colour finishing. Furthermore, a textured surface on a die casting with large flat areas overcomes problems of flow marks by breaking up the surface. Selective painting on parts of the plated surface can be achieved by spray painting through a mask moulded to fit the surface closely or, alternatively, if the areas to be left unpainted are raised, the whole casting is painted after plating and the raised areas are polished with an abrasive hard enough to remove the paint but not damage the plating.

Metallizing

Vacuum metallizing is a process for applying a thin metallic film onto a prepared lacquered surface within a high vacuum. Many metals can be metallized but aluminium is commonly used. The appearance of copper, silver, brass and gold can be simulated. Until recently the process was employed for coating plastic mouldings but metallizing has now been extended to zinc alloy die castings[8]. Parts to be treated do not need polishing if surface defects can be levelled with a base coat of lacquer applied after degreasing. Following a low temperature stoving treatment, a thin film of aluminium is vaporized onto the parts under vacuum. A second application of lacquer followed by stoving leaves a bright silvery finish which can be dyed in any colour.

Plastic coatings

Sprayed epoxy powder coatings can be used in place of paints to give a tough, adherent and corrosion resistant coating. The degreased castings, preferably chromate treated, are electrostatically sprayed with a powder paint and stoved for a few minutes to cure the coating. The technique avoids the hazards of inflammable thinners used with conventional paints and it is easy to change the powder reservoirs to apply a fresh colour. Thick plastic coatings can be applied to degreased zinc die castings by heating them to about 300°C followed by dipping into a fluidized bed of low melting point nylon particles. The nylon forms a fused layer which solidifies on cooling. Sharp edges and corners must be avoided and, as the coating is about 0.4 mm thick, any fine detail on the surface becomes blurred.

Anodizing for zinc alloys

A corrosion-resistant zinc anodic coating process was introduced in 1957, following development work under the sponsorship of the International Lead Zinc Research Organization[9]. The process involves anodizing zinc alloy at a potential of up to 200 volts in a solution containing ammonium phosphate, chromate and fluoride. As formed, the casting is porous but can be sealed in a hot dilute sodium silicate solution or an organic paint film. Anodic coatings, ranging in colour from olive green, light grey, to charcoal grey, with thicknesses approaching 30μm, have withstood over 1500 hours exposure to a 5% salt spray accelerated corrosion test before any signs of corrosion appeared. Good resistance to detergent solutions is demonstrated in zinc alloy washing machine components that have been anodized. A paper by K. Wright[10] describes how anodizing overcame a problem associated with die cast valve plates in an automobile windscreen wiper motor. Failure of the as—cast plates had been occurring after about 70 000 cycles, but following anodizing life test requirements of 7 million cycles were being exceeded; some plates have been enduring 12.5 million cycles with no measurable wear.

Other finishes for zinc

There has been much interest in lacquer formulations to preserve the bright polished surface of untreated zinc alloy castings. Polyurethane lacquers containing rubeanic acid have been developed[11] to overcome problems of yellowing on exposure to light and dulling due to moisture penetration. Acrylic lacquers have also been used successfully with rubeanic acid additions, but a stoving treatment is required to cure the surface. Some technical difficulties remain to be overcome in applying the lacquer without adding unduly to finishing costs.

The Zincart process, developed in Japan by the Mitsui Metal Arts Company, is a surface chromating treatment but with a chloride addition made to the etching solution. Colour is introduced with dyes followed by a protective acrylic lacquer covering. Several colours are available but only the matt black finish has achieved extensive application in Japan.

Development and trade associations in many countries publish useful brochures with instructions for the many finishes which can be applied; for example in Britain two publications[12,13] are issued jointly by the Zinc Development Association and the Zinc Alloy Die Casters Association.

Finishes for aluminium alloy die castings

The thin film of oxide on the surface of aluminium alloys affords some protection against corrosive attack, and for many applications further treatment is not required. Mechanical processes, for example tumbling castings in a barrel with selected burnishing media and polishing compounds, provide an acceptable surface finish. Buffing will achieve smoothness and reflectivity on the casting surface, but the final effect will depend to some extent on the alloy and the hardness of the casting. Matt surfaces can be produced by shot blasting and satin smooth surfaces by wire brushing. Treatments which offer improved protection from corrosion and attractive surface finish include painting, electroplating and anodizing.

Various chemical dip processes are suitable for producing decorative and corrosion resistant finishes on aluminium alloy die castings which are to be painted or left unpainted. Where abrasion resistance is needed, the chemical treatment alone is not sufficiently protective but is used as a pretreatment for painting. Alocrom 1200 is a proprietary non-electrolytic chemical dip process developed by ICI and approved to DTD 900/4413A. An aqueous solution of Alocrom powder, with small additions of nitric acid, is held in a container of stainless steel, plastic, synthetic rubber or other acid resistant material. After degreasing, castings are immersed for 2–5 minutes in the bath held at about 25°C and a golden yellow coating forms. Excess chemical is removed from the surface by flushing with clean water followed by drying in air or wiping with cloth. If the castings are to be painted, it should be done as soon as possible to minimize contamination. Aluminium die castings can be painted, enamelled or lacquered as readily as any other materials following the application of a suitable primer.

Electrodeposition on aluminium alloys

For many years it has been possible to electroplate aluminium, but the protective film tended to become separated from the base metal under conditions of stress or vibration. Since aluminium is amphoteric, it must be protected against the attack of acid or alkaline plating solutions by the application of an adherent and chemically resistant deposit of a more cathodic metal. The development of the Zincate immersion process for pretreatment of aluminium, followed by plating with nickel or any other metal provided a surface which was sufficiently adherent to give a metallurgically sound bond. However, when this technology was extended to the treatment of alloys it did not prove very satisfactory, owing to the lack of control of the immersion film thickness and structure. Studies of the film characteristics[14,15,16] led to the development by W. Canning Materials Ltd. of zinc alloy immersion deposition known under the trade name of Bondal process. This uses a single immersion solution to deposit an interply 0.0001 mm thick consisting of 85% zinc, 10% copper, 4% nickel and 1% iron to be used as an undercoat for deposition of any commercially used metal.

Although originally developed mainly for pure aluminium to be plated for decorative purposes, the Bondal process was quickly applied to a range of aluminium alloy castings, forgings, pressing and extrusions. Since the physical and chemical properties of the composite depends on the characteristics of the bond, these were extensively studied and an optimum processing cycle has been developed, suitable for all the requirements.

The most difficult problem to be overcome in obtaining satisfactory plating of aluminium alloys has been the selection of the etch solution to remove the non-metallic debris from the surface of the component to be plated. Dilute nitric acid had been found suitable for commercial purity aluminium, but several alloys demanded a more flexible approach. Mixed nitric, sulphuric, hydrofluoric and chromic acids are used, with the addition of ferric chloride, fluoride or hydrogen peroxide, as required in individual cases. The etching process used often predetermines the quality and bond strength of the deposit. The adhesion of the plated deposit was found to be a combined function of its thickness, crystal structure and chemical composition which, in turn, determined the current density available for the extent of nucleation and coverage of the alloy by the deposit. The method by which sound deposits could be produced was the effective control of the growth rate through properly designed growth inhibition.

Anodizing

A surface oxide film developed under closely controlled conditions by the electro-chemical anodizing process is effective in increasing the corrosion resistance of aluminium and its alloys. Surface properties achieved vary with the type of alloy and the composition and operating conditions of the electrolyte. With high purity aluminium and certain aluminium alloys, an attractive silvery coloured reflective anodic finish is achieved which can be coloured using pigmented salts. To achieve acceptable surface appearance, castings should be free from flow marks and polishing defects and the care which must be taken in casting preparation before anodizing represents a significant part of treatment costs. Surfaces can be polished using multistage buffing, while satin and matt finishes are produced using brushes or abrasive blasting. Before anodizing, all parts require degreasing followed by chemical or electrochemical brightening or polishing.

Sulphuric anodizing processes, used for both decorative and protective applications, are capable of producing a wide range of film thickness, hardness and other properties. Alternatively oxalic acid, used to a limited extent in Europe, provides light shades of gold or bronze without dyeing. Chromic acid processes, although producing thinner and less wear-resistant films than sulphuric acid, offer better corrosion resistance for equal film thickness.

Parts to be anodized are attached to jigs or racks which become the anodes in the electrolytic cell; separate cathodes of lead, aluminium or stainless steel are used. Depending on the type of electrolyte, the bath operates at a potential of $10-20$ volts within a temperature range of $18-25^\circ C$; the lower the temperature the harder and less absorbent the anodic film. Surface build up is controlled by the alloy composition, current density and duration of treatment. For indoor applications, thicknesses from $0.005-0.015$ mm are suitable; but for severe outdoor service a thickness of about 0.03 mm is necessary.

A thin oxide barrier film is produced on the base metal surface with a thicker porous outer layer consisting of columnar cells containing open pores. For maximum corrosion resistance, the porous layer must be sealed after anodizing, by immersion in boiling water, producing a surface oxide swelling that constricts and finally closes the pores. Colour is usually introduced by incorporating pigments in the porous layer before sealing; alternatively, oxides of alloying elements in the aluminium give the desired colour effect during anodizing. With this integral colouring method it is more difficult to achieve a uniform colour, due to segregation of alloying elements and non—uniformity of grain size which affects surface appearance.

The composition of an aluminium alloy has a considerable influence on the protective properties and colour of anodic films, and affects the resistance of the film to corrosion. Oxide films produced by anodic treatment on pure aluminium are highly reflective and silvery in appearance. *Table 29.1* indicates the effect of secondary elements in aluminium on the colour of the anodic film produced.

TABLE 29.1 Some physical properties of oxides of metals frequently present in aluminium alloys

Oxide	Colour	Refractive Index
Al_2O_3	White/Transparent	1.65
SiO_2	Transparent	1.43–1.57
MgO	White	1.74
ZnO	White	2.00
$Fe_2O_3H_2O$	Yellow	2.05–2.31
Mn_3O_4	Black	2.15–2.46
TiO_2	White	2.50
Cr_2O_3	Green	2.50
CuO	Black	2.63
Cu_2O	Brown	2.70
Fe_2O_3	Red/Brown	2.78–3.01

Comparing the refractive index and colour of metal oxides whose elements are commonly present in die casting alloys, either as impurities or alloying elements, it is shown that magnesium oxide, with a similar refractive index and colour to aluminium oxide, is suited for alloying with aluminium for subsequent anodizing. LM5 alloy with 3.0–6.0% magnesium can be anodized successfully to give a transparent colourless to whitish film.

The common pressure die cast alloys are rich in silicon and, although a protective anodic finish can be produced, it is grey in colour. This is associated with the fact that silicon is virtually unaffected by the usual anodizing techniques and remains entrapped in the anodic films, causing them to become darkened. Such films are only suitable for pigmenting with darker colours. The majority of other constituents are usually attacked and, by direct anodic dissolution or the formation of an oxide which is soluble in the electrolytes, are likely to dissolve partially or fully during the anodizing process.

Finishes for magnesium alloy die castings

Magnesium alloys are rarely used in the as cast condition and the minimum treatment consists of applying a corrosion inhibitive film to the casting. Organic coatings are often specified following pretreatment with a corrosion inhibitor.

The standard protective treatment for magnesium, to prevent reaction with moisture during storage or transit or to provide a suitable base for painting, is a chromate dip process to specification DTD 911. When vibratory barrelling or abrasive blasting is used to clean castings and remove flash and sharp edges, ferrous media should be avoided, since they become embedded in the casting surface, leading to localized corrosive attack. The preferred medium is either bonded ceramic chips for tumbling and vibratory barrelling, or fine aluminium or glass beads for blast finishing. After a final chemical clean of the parts, either by an emulsifier, an alkaline dip or solvent cleaner, they are immersed for about 30 minutes in a boiling solution of water containing ammonium sulphate and ammonium and potassium dichromate contained in a mild steel or, preferably, an aluminium vessel. The pH value of the solution is maintained at about 6.0 for the production of a film of good appearance with additions of ammonia, chromic acid or sulphuric acid as required to replace loss by evaporation. The colour of the film on machined surfaces will vary from brown-black to grey-black when the content of aluminium in the alloy increases over 4%. After treatment the parts are washed in warm water or a dewatering oil dip containing lanolin.

Chromating provides a suitable base for subsequent organic finishes. For painting, the parts are either degreased to remove any oil film or painted after chromating and drying, so eliminating the need for an oil dip. Priming with zinc chromate paint followed by an undercoat containing flake aluminium provides a surface suitable for any paint which is compatible with the primed and undercoated surface. Mercury or lead-based pigments must be avoided in the coating, since in the presence of moisture these metals can be precipitated electrically, causing localized cell corrosion.

Anodic coatings for magnesium

Anodizing provides a surface coating to magnesium alloys that offers good resistance to corrosion and wear and which readily accepts decorative paint finishes. Two well known techniques, the DOW 17 and HAE treatments, have been recognised as offering protection to magnesium in corrosive environments, while the MGZ anodizing process[17] invented by H.A. Evangelides is reported to be capable of producing strong wear resistant coatings in short processing times.

Components to be treated are made the electrodes in an aqueous solution of chromate, vanadate, phosphate and fluoride salts. An alternating current is applied to the bath at a relatively high voltage. The adherent ceramic like coating consists of mixed glassy and crystalline phases of the oxides and fluorides of magnesium, chromium, vanadium and phosphorus. Varying treatment times will produce coating thicknesses of 0.01–0.03 mm, dark green to black in colour, depending on the alloy and electrolyte composition. Development of this process is continuing, the main applications being on moving parts requiring high wear resistance.

Plating of magnesium die castings

Although methods for plating magnesium were known in the early 1950s, their use was limited to aerospace and telecommunication equipment and they were not suitable for the commercial finishing of die castings. Later development work led to a greater understanding of how magnesium alloy composition and casting conditions affected platability. The method now used is similar to that described for aluminium plating in that both metals need to be coated with a thin layer of zinc before nickel plating, but different formulations are required for magnesium. The process includes surface conditioning, cleaning and activating, followed by an immersion zincating treatment.

Although die castings in AZ 91 can be plated, problems arise because areas of the aluminium-rich intermetallic compound $Mg_{17}Al_{12}$ tend to segregate and adversely affect the precipitation mechanism of the zinc. When only about 6% aluminium is present, as in in AZ 61, the segregation is much less. The 7% aluminium alloy AZ 71 is also suitable for plating and is easier to die cast than AZ 61. It has been reported[18] that the alloy known as ZA 124, mentioned on page 140, has a combination of fluidity and platability.

The procedures for plating magnesium were described in a paper[19] by A.L. Olsen, given to the Institute of Metal Finishing in 1980. After cleaning, the surface is activated with an acid treatment, followed by rinsing and a secondary alkaline activation. The zincating bath consists of zinc sulphate and alkali metal pyrophosphate with an alkali metal fluoride to control the rate of deposition. Sodium carbonate is added to bring the pH value to 10.2–10.5. Zinc is deposited on the magnesium alloy surface; the treatment time is about 2 minutes at 50–65°C.

Zinc, cadmium, copper and brass can be applied directly on the zincate coating from standard cyanide baths. If the parts are to be plated subsequently by bright nickel, copper deposits with a minimum thickness of $7\,\mu m$ are required (comparable with the thickness of copper usually applied to zinc die castings) to prevent chemical attack on the base metal. When plating deeply recessed parts it is necessary to increase the thickness of copper. After this treatment any further plated deposits can be applied with normal procedures. Plated magnesium alloy die castings will meet the requirements of a number of uses, for furniture, household appliances and parts used in building. Tests are being conducted on automobile hardware and it is considered that multilayer plating systems including microdiscontinuous chromium may provide the necessary protection for outside door handles. In Olsen's paper, referred to above, illustrations are shown of magnesium alloy die castings with hard chromium deposit, zinc plating, gold plating and a variety of nickel–chromium plate finishes on car door handles.

References

1.   Chromate passivation of cadmium and zinc. *UK Defence specification DEF 130*

2.   BAUDRAND, D.
     Zinc plating on zinc die castings. *Society of Die casting Engineers* Congress Paper
     G-T77-023 (1977)

3.   SEYB, E.J.; JOHNSON, A.A.; TUOMELLO, A.C.
     Proceedings of American Electroplaters Society 44 p. 29 (1957)

4.   *French patent* 1 447 970
     British patent 1 122 295 and 1 187 843

5.   DAVIES, G.R.
     A simple treatment for improving corrosion protection given by nickel–chromium.
     Electroplating Metal Finishing, **21,** 393 (1968)

6.   CARTER, V.E.; CHRISTIE, I.R.A.
     The outdoor exposure performance of chromium electrodeposited from a trivalent electrolyte.
     Transactions of the Institute of Metal Finishing, **51,** 41 (1973)

7.   NICOLAI, J.R.; MARCE, R.E.
     Better service from zinc die castings in mechanical applications. *Society of Die Casting Engineers*
     Congress Paper G-T79-085 (1979)

8.   GABOWER, J.F.
     Vacuum metallizing on zinc and aluminium castings. *Society of Die Casting Engineers* Congress
     paper G-T79-081 (1979)

9.   *US Patent* 3 011 958
     Canadian patent 605 265
     UK patent 876 127

10. WRIGHT, K.
     Zinc anodizing at Trico-Folberth. *Ninth International Pressure Die Casting Conference, London.*
     Zinc Development Association, London (1978)

11. CHRISTIE, I.R.A.; CARTER, V.E.
     Performance of lacquers for copper and zinc. *Transaction of the Institute of Metal Finishing,*
     50, (1), 19 (1972)

12. *Zinc Die Casting Guide,* 3rd Edition. Zinc Alloy Die Casters Association, London (1978)

13. *Finishes for zinc alloy die casting.* Zinc Alloy Die Casters Association, London (March 1979)

14. WYSZYNSKI, A.E.
     An immersion alloy pre–treatment for electroplating on aluminium. *Transactions of the
     Institute of Metal Finishing,* **45,** 147 (1967)

15. WYSZYNSKI, A.E.
     Electrodeposition on aluminium alloys. *Transactions of the Institute of Metal Finishing,* **58,** 34
     (1980)

16. GOLBY, J.W., DENNIS, J.K.; WYSZYNSKI, A.E.
     Factors influencing the growth of zinc immersion deposits on aluminium alloys. *Transactions of
     the Institute of Metal Finishing,* **59,** 1 (1981)

17. KOTLER, G.R.; HAWKE, D.L.; AQUA, E.N.
     A new anodic coating process for magnesium and its alloys. *Society of Die Casting Engineers*
     Congress paper G-T77-022 (1977)

18. FOERSTER, G.S., GALLAGHER, P.C.J., HAWKE, D.L.; AQUA, E.N.
     Research in magnesium die casting. *Die Casting Engineer,* **21,** (1), 12 (January/February 1977)

19. OLSEN, A.L.
     Plating of magnesium high pressure die castings *Transaction of the Institute of Metal Finishing,*
     58, 29 (1980)

Appendix : Units of Measurement

Metric and SI units have been used wherever possible, but conversions to Imperial units are indicated in various places, since the older units are still quoted in many publications.

Temperature

All temperature measurements are expressed in degrees Celsius (Centigrade). The conversions shown in *Table A1.1* may be helpful.

TABLE A1.1

	Temperature (°F)	Temperature (°C)
Melting point of zinc alloy	729	387
Melting point of aluminium alloy LM24 (380.0)	1076	580
Melting point of brass	1652	900
Pearlite eutectoid point in steel	1350	732

Dimensions, and limits of accuracy

One inch equals approximately 25.4 mm. In converting limits of accuracy from inches to metric, the limit expressed in inches is multiplied by 25.4, omitting the last figure and rouding off to the nearest unit. Thus ± 0.005 in is first converted to plus or minus 0.127 mm and then rounded off at 0.13 mm. BS 2858, 1973 gives information about conversions of inch and metric dimensions on engineering drawings.

Weight

One pound equals approximately 0.454 kg. One ton (2240 pounds) equals 1016.96 kg. The standard metric measure of tonnes (1000 kg) has been used, except for circumstances where the use of 'tons' seems more appropriate, as for example in the description of the first American production of die cast engine blocks.

Machine locking force

The units of tons and tonnes are so nearly alike that there is no point in converting from one to the other. 'Tonnes' has been used in descriptions of modern machines but the locking forces of older machines have been stated in tons.

Mechanical properties

The pound force per square inch, shown by the abbreviation psi, is still used in America and Britain, but the SI nomenclature for tensile strengths is being adopted to an increasing extent. Tensile strengths are expressed in Newtons per square mm [N/mm^2.] The following conversions are used:

One ton force per square inch equals 15.44 N/mm^2
1000 lb force per square inch equals 6.89 N/mm^2

Impact force measurements which previously were expressed in foot-pounds force are stated in Joules, with the conversion 10 ft lb [force] = 13.56 Joules.

Proof stress

In the past, proof stresses have been expressed as the amount of stress required to cause a permanent elongation of 0.1%; this is similar to the elastic limit of a material and easier to measure. More recently, proof stress measurements of 0.2% have been used. The following figures indicate the differences between the two *(Table A1.2)*.

TABLE A1.2

| | Stress in N/mm^2 | |
	LM24 test bar	*Mild Steel*
Ultimate tensile strength	190	540
0.1% proof stress	90	300
0.2% proof stress	110	340

Pressures

The SI unit of pressure is the Pascal (Pa), which is one Newton per square metre (N/m^2). As it is difficult to visualize a pressure spread over a metre in some engineering situations, including die casting, pressures are often expressed in Newtons per square mm (N/mm^2) or, for large pressures, kilo Newtons per square mm (kN/mm^2). The following conversions may illustrate the relationship between old and SI units *(Table A1.3)*.

TABLE A1.3

	Imperial	*Pascals*	*SI Newtons*
A casting pressure test	30 lb/in^2	0.2×10^{-6}	0.2 N/mm^2
Pressure of a compressor	150 lb/in^2	1.034×10^{-6}	1.034 N/mm^2
Injection pressure of a die casting	2 $tons/in^2$	30.9×10^{-6}	30.9 N/mm^2
Gas cylinder pressure	2000 lb/in^2	13.7×10^{-6}	13.7 N/mm^2

The c.g.s. pressure unit of one Bar equals 100 000 Newtons per square metre, or 0.01 Newtons per square millimetre.

Hardness measurements

The Brinell hardness tester uses a steel ball pressed into the surface of the metal; for measuring soft metals a 500 kg load is applied and for steels a 3000 kg load. The Brinell test is not a satisfactory measure for harder materials such as alloy steels. The Rockwell tester is a direct indentation machine; to give maximum accuracy, a number of scales, from A to G, are used and within the groups different loads are applied, as shown in *Table A1.4.* The materials mostly used in die casting are measured on the A or C scales, the latter being applied for the hardness of hot die steels.

TABLE A1.4

Rockwell scales	Penetrator	Load (kg)
A	Diamond	60
B	1/16 in (1.6 mm approx) diameter ball	100
C	Diamond	150
D	Diamond	100
E	1/8 in (3.2 mm approx) diameter ball	100
F	1/16 in diameter ball	60
G	1/16 in diameter ball	150

The Vickers hardness tester is not so widely used as Brinell or Rockwell on account of the careful surface preparation that is required and the need for great precision in the optical determination of indentation size. However, this method has the advantage over Brinell and Rockwell that no change in indenter or load is necessary for the measurement of either soft or hard materials. *Table A1.5* shows some typical metals used in the die casting industry, with their Brinell and Vickers hardnesses, although the correlation of one scale to another can be only approximate. The final column shows conversions to appropriate Rockwell scales, but in each case the Rockwell scale most likely to be selected is indicated by the figure appearing in a square.

TABLE A1.5 Hardness scales

Metal	Brinell 10 mm ball 500 kg load (*3000 kg load)	Vickers	Rockwell						
			A	B	C	D	E	F	G
Aluminium alloy LM 6	60	65	[21]	17			67	66	
Aluminium alloy LM24	85	96	[36]	52			88	86	6
Zinc alloy BS1004A	83	93	[35]	50			87	85	3
Zinc alloy BS1004B	92	104	[39]	58			92	90	15
40 ton steel	170	176	[58]	86	6		108	106	59
2% Beryllium–copper	370*	394	[70]		40	55			
5% Chromium steel	485*	513	76		[50]	63			
Nitriding steel	780*	1170	87		[70]	79			

Alloy specifications

Although efforts are being made to persuade users of casting alloys to employ international standards, the long established use of 'LM24' or '380' is so simple and convenient that works personnel and others can hardly be blamed for resisting the use of 'AlSi8Cu3Fe'. Nevertheless it is desirable that in due course there should be unification of specifications. The problem is further complicated because different countries require different limits of composition. For example the varied designations for the aluminium—silicon—copper alloy LM24 is discussed on page 29. Where it appeared to be helpful that the specifications in different countries should be compared, suitable Tables have been included.

Index